K.C. -
Nice to see you Do.
Hope you enjoy the
humor & predator
predicaments.
Bonn R. Wyler
2/17/18

ADVENTURES IN THE BUSH: AFRICA TO ALASKA

Bruce K. Wylie

Bruce K. Wylie

Adventures in the Bush: Africa to Alaska

Print ISBN: 978-1-54390-385-0

eBook ISBN: 978-1-54390-386-7

© 2017. All rights reserved. No part of this publication may be reproduced, distributed, or transmitted in any form or by any means, including photocopying, recording, or other electronic or mechanical methods, without the prior written permission of the publisher, except in the case of brief quotations embodied in critical reviews and certain other non-commercial uses permitted by copyright law.

ACKNOWLEDGEMENTS

Thanks to my wife, Mariama, for her patience with my love of the outdoors and adventure, as well as my crude social skills. Thanks to the role models provided by Dad and the late Moukhamadoune Abdurrabi on curiosity, commitment to academia and science, compassion for the poor, and expressions of the joys of life.

DISCLAIMER

This book is a compilation of true short stories or campfire stories which captures my recollection of events. Everyone knows my memory is perfect! The stories are arranged in rough chronological order. I used mostly only first names hoping that this will allow folks to claim or disclaim notoriety associated with this book. The comments and generalizations made in this book are entirely mine and completely independent from any current or prior employers of mine.

FOREWORD

Adventures in the Bush is the story of a man with cowboy spirit in his blood, forever home in the outdoors, always striding forward into the next adventure. Not much ruffles Bruce Wylie's feathers. Not frozen toes or a trek across the Sahara or the unusual courtship rituals of Niger. Not a dinner of "thousand-year-old eggs." Not even a charging grizzly bear. He faces life's challenges with problem-solving skills nurtured by parents who allowed him and pushed him to spread his wings. Travel with Bruce from his rural American roots to the remote reaches of Africa, to the stark plains of Asia, and to the frozen tundra of Alaska. You'll be amazed and entertained, fascinated and inspired by an unflappable lover of life.

I met Bruce thirteen years ago when I accepted the position of librarian at the U.S. Geological Survey Earth Resources and Observation (EROS) Center in Sioux Falls, South Dakota. USGS EROS is a satellite data and science center that monitors and assesses changes in the Earth's land areas. Bruce is a senior scientist at the center, and an internationally recognized scientist in carbon research. As librarian, I've had the pleasure, and honor, of working with Bruce, supporting his library research needs. He's also been very generous and allowed me to use him and his vast body of published work as my "test case" when assessing various online information tools.

But I've also been fortunate to get to know Bruce beyond his professional life. Bruce is a member of the EROS Toastmasters Club— a charter member, in fact. Over the years, Bruce has regaled our club with story after story of his life's adventures. Adventures around his early life in Wyoming and Nebraska and Montana; around his Peace Corps and professional activities in Africa; and around his fieldwork as a research scientist in Alaska and Asia. Adventures around his hobbies: kayaking, motorcycling, hiking, birding, camping, and skiing. Adventures around one of his many volunteer activities: dog-walking. Just about anything that will get Bruce outdoors is on his to-do list, and he tackles it all with gusto. Curious, generous, undaunted, Bruce Wylie embraces life as we all should: as a gift to be opened and caressed and appreciated each and every day.

Carol Deering, Librarian
Don Lee Kulow Memorial Library
Innovate!, Inc. | Contractor, U.S. Geological Survey (USGS) Earth Resources Observation and Science (EROS)

— 1 —

COWBOY (1961-1974)

In my early years in southwestern Wyoming (Mountain View, 3 years old through 3rd grade), all I wanted to be when I grew up was a cowboy. I remember when Mom and Dad finally broke down and bought a TV. My brothers and I—I do not think my sisters were old enough to even comprehend—had no idea what a TV even was. It was a black and white model, and my brothers and I were soon mesmerized by Roy Rogers and *The Lone Ranger*. We would sit astride the arms of the couch, and during a chase scene we would kick and whip the couch arm frantically like it was a horse. We had strict limits on how much TV we could watch. By fourth grade we had moved to western Montana (Seeley Lake) where I read a book called *A Horse for the Winter*. It was about a girl whose father knew she wanted a horse. He was able to get her a horse to take care of in the winter and return to its owner in the spring. This was a novel idea, and I am pretty sure that I made sure Dad was familiar with the book. After a while, I kind of resigned myself to living without a horse. Several years later Dad surprised me by arranging for us to get two horses (a white mare and he colt) on loan for the winter from a member of the church, the Grays.

It must have been fall when we went to get the horses. The horses were kept in a pasture behind a closed saw mill. The owner, Marcia Gray, was having some trouble catching the horses, particularly the colt that was to come with us. Marcia and I were watching the young foal running around as we tried to round up the horses with a pickup. I asked her what the name of the colt was. Marcia replied that I could name the colt since it had no name. I commented that it sure was a cute little bugger. I paused and then I said, "That's her name! Bugger!" Later I found out that Marcia complained to my Dad about the name. Apparently, Bugger had an obscene derogatory meaning. Dad argued, however, that Bugger, in this context, was a harmless word. So, the name stuck.

The first obstacle was getting the mare and foal to our place, which was down about 7–10 miles of road, mostly gravel.. Dad's plan was that I was going to ride the mare while he followed in the car, assuming the foal would follow. Dad guided me on which of the roads I needed to take. On a remote gravel road along the west side of Seeley Lake, we came to a cattle guard. Dad went ahead and opened the gate on the edge of the cattle guard. I rode the mare, Snowy, through the gate and kept on going at a walk with the mare, trying to entice that crazy foal, who was half wild and not too familiar with humans, to come through the gate. Instead of going through the gate, Bugger was so afraid of Dad standing there holding the gate open, that she ran across the cattle guard! Dad and I were both shocked and stared at each other. I was sure Bugger would break a leg or two. Her back two legs had fallen between the rails of the cattle guard, but somehow, Bugger got across the cattle guard without breaking any legs by doing some really high stepping.

I remember it being a long way home, and I was pretty tired, but there was no way I was going to complain! Previously Dad had my brothers and I help build a corral behind our garage. We used two-by-fours from a stud mill donated by another member of the church. The posts were just a two-by-four with a shorter angled leg on the lower third of the post, kind of like an upside down "Y." I did not think our fence would even stand, but as we nailed the cross boards from post to post, the structure stiffened and my confidence in it increased. The fence poles leaned slightly outward from the interior of the corral, with the second leg of the pole extending outward and downward. In the 3 years or so that we had Bugger and Snowy, our fence only fell over once on a real windy day. Dad also cut a door in the back of the garage into what used to be a woodshed. This allowed the horses to get out of the rain or cold, but even though there were tracks and manure in there, I rarely saw them in the woodshed.

I must have been in 5th grade or so, as the only way I could climb on the mare was to lead her up to a stump or rock for me to stand on and jump from there. Snowy was an older, savvy mare, and she often would resist standing next to an object that I could climb on board from. Once while I was riding her, she just stopped on a trail, out of reach on any branches I could have used for a whip and where there was no stump or other object for me to remount from. I just kept kicking and yelling for 20 minutes or more before Snowy gave up and proceeded. I had won the war of stubbornness!

It was my job to feed the horses every morning and make sure they had water. Initially I told Dad I should probably get up earlier than the rest of the kids so I could feed the horses before school. Dad said, "No, you just need to be more efficient with your time in the morning." He was right. I just popped out of bed every morning, got dressed, and stepped out and hollered at the horses and they would whinny back at me as I headed over to toss part of a bale into their feed bunk. In the evening, after school, I would go riding on Snowy with Bugger following behind. I did a lot of exploring and developed a comfort being alone out in the forest in new country. I had a network of trails that departed through the woods behind our back yard and along the river, which was 3 blocks up the street.

Soon I noticed a book around the house that had mysteriously appeared, *Breaking and Training the Stock Horse*. I was curious and found it interesting and useful. It promoted a gentle approach to "breaking" a horse, which Dad said he liked, rather than a strong-arm approach. I think this horse breaking logic stuck with me through the rest of my life—I applied these horse breaking approaches to people and social situations.

Our neighbor had a small round corral. Dad must have talked to him about helping break Bugger. Initially she was so wild, no one could get close to her. I was told to ride Snowy into the round corral and Bugger followed. We removed Snowy and then Bugger freaked out when she was alone in the coral. The neighbor lassoed Bugger and Bugger tugged and pulled and fought the rope. The neighbor gave ground to Bugger but kept constant tension on the loop around Bugger's neck. Finally, Bugger was running out of air and fainted. The neighbor ran up and loosened the loop so Bugger could breathe and stepped back to the other end of the rope. Soon Bugger started waking up, stood up, and just stood there. The neighbor slowly approached using a slow, smooth tone of voice and worked up close enough to Bugger to pet her. She was startled but just stood there. He then started leading her around a bit and she followed.

Bugger's training was progressing, and in the second winter I was trying to get her used to something on her back. I was using a saddle blanket, rubbing it on her and sliding it across her back. I was in the corral, and with spring approaching, there was a mixture of melting snow, manure, and urine on the ground in the corral. Bugger did not like the blanket on her back, but I persisted, stepping alongside as she tried to move away. To be clear, there was no halter or lead rope on Bugger. In a sixteenth of a blink of an eye, Bugger kicked me in the stomach with her hind hoof. I spun around partly to dodge, but I was about 3 seconds too late. I fell face first in the melting snow and manure. I stood up and walked to the house as fast as I could, but I could not breathe. When I got to the backdoor and opened it, to my surprise Mom was right there! Mom looked at me and would not let me in the house. I whispered that I could not breathe and her response was, "Go lay in the snow bank over there!"

Somehow laying on my back arching slightly backward on that snow bank, I started to breathe. The air in my lungs felt absolutely divine, despite the taste of horse manure and urine in my mouth and nose. I was relieved and happy to still be alive! Mom soon returned with some rags and warm water as well as a bathrobe. I changed out of my filthy clothes outside on the cement back step into the robe and went into the house to take a shower. I had survived!!

Later in the early summer, Dad came to me with a glass of water. He said, "Taste this. Does it taste like horse pee?" I guess I was the resident expert on what horse pee tastes like. I took a drink and said, "Yes." Dad put some Clorox down into our well and soon after the taste in our well water dissipated. The spring melt was over and there was reasonable flow gradient in our ground water, so maybe the improved water was due not to the Clorox but to nature taking its course?

In the early days of training Bugger for riding, my older brother Allan would often ride Snowy while I rode Bugger. I was gentle with Bugger and she insisted on a long, loose rein and often ran with her head very close to the ground. This is a very vulnerable position for the rider, but I assumed she needed to be sure of her footing because of my additional weight. Allan and I were trying to get Bugger to run faster. She just kind of ambled along. I told Allan to gallop on ahead with Snowy up around a slight bend in the grassy remote road we were on. The plan was that this would bait Bugger to run a little faster. That is the last thing I remember.

I woke up lying on the ground with Allan trying to put his jacket under my head as a pillow. I did not remember riding Bugger and was incredulous that I had even been on her back. Allan said he was galloping along on Snowy when suddenly Bugger caught up to him without a rider. Bugger was nearby eating grass trying to look innocent, but she was still keeping an eye on me. I had a big bloody cauliflower ear, and when we got home Mom and Dad stayed up with me into the night. I felt nauseous but they did not want me to vomit. I think they thought vomiting might mean that I had a concussion. In the end I finally did vomit, felt better, and then fell asleep. My memory of recent events gradually returned over the next week or two. I think it took me two to three weeks before I again began riding Bugger. I suspect this also made Mom and Dad happy, as a repeat concussion before recovery can be serious.

I had the horses for about three winters and some summers. I spent a lot of time exploring the woods and national forest lands around our house. We moved to the Sandhills in Nebraska when I finished 6th grade. In Stuart, Nebraska, Dad soon had me lined up riding various people's horses and ultimately riding horses that had bad habits in need of correction. I continued riding horses pretty much through high school.

— 2 —

FIREFIGHTER (1974-1978)

I was feeling a little grouchy, logy, and listless, which was typical for me when I had to be up and out the door at pre-dawn. It was early summer in the northern Sandhills of Nebraska as I started my 1966 VW bug and headed for highway 20 and pointed westward. I had recently graduated high school in Stuart, Nebraska, fifth in my class of 30 or so. It was 1974 and I had somehow landed a summer job on a Forest Service brush crew at the 9 Mile Ranger District, west of Missoula, Montana.

As I left Stuart, I came to a major turning point in my life. It was the township gravel road, heading south to my buddy Terry Murphy's family ranch, just west of Stuart. I had worked with Terry's family on numerous occasions on cattle drives and branding and vaccination of the calves. I knew the older Murphy kids well (there were 11 of them). These cattle drives were in the summer. After branding and vaccinating the calves, we took the cattle to the Murphy summer ranch way down south in the Sandhills. Cornelius Murphy, Terry's dad, had a plane which he used to check the cattle in the summer. The second cattle drive was in January in the dead of winter. We brought the cattle up to the Murphy home place near Stuart, where they would be fed through the winter and calve in the spring. What was tugging at my stomach that morning was whether to abort this trip and summer job in Montana and play it safe and stay in Stuart. Another complication was that I had been dating Terry's younger sister. I had also developed a network of friends since my Dad, Harold, moved our family there 6 years earlier. It was a tough decision—go back to the comforts of home, or go west to the unknowns in Montana? I pulled over at the intersection to contemplate.

I wanted to be the first of my siblings to leave the "nest" at Stuart. My older brother, Allan, had backed out of going off to college at the last minute the previous year to work at a local feed lot. I knew Dad wanted all of his five children to attend college. That was drilled into all of us from early grade school on. Dad had encouraged me to apply to Prescott College, an outward bound, backpacking, and a "no grades" school in Arizona, and I had been accepted. I finally pulled out on to Highway 20 and headed west. I drove slowly for a bit, still pondering my decision. I figured I was ready to get out into the world and learn new things and meet new people. Besides, if things went south, I could always "high tail it" home. I drove faster as the supporting arguments for "westward ho" where gaining momentum in my head. Finally, after about a mile, I made it up to the 55-mph speed limit. I was on my way to 9 Mile! The die was cast for me to test my fate, skills, and character beyond Stuart and Mom and Dad's protective sphere of influence, or so I naively thought.

My route took me westward to Spearfish, South Dakota, and then north up to Highway 212, and west into Montana. One of first towns I came to in Montana was the tiny town of Alzada. I wondered what the drinking age was and pulled in to the little bar/cafe there (there was not much more to Alzada than that). I ordered a breakfast of pancakes and eggs, my favorite. As the middle aged, slightly overweight, Latino waitress brought out my food, I asked about the drinking age. "Eighteen," she said. I made some comment about me being legal to drink in Montana. The waitress asked if I wanted a beer. "No, I would pass," I said, as I could now get a beer any time (I was going to be in Montana all summer after all), and I had a long, solo drive across the state ahead of me.

By mid-morning the next day, I was heading from Albertson, Montana, toward 9 Mile. I turned up the gravel road to the 9 Mile Ranger Station. There was no town of 9 Mile, just pastures, trees, and what appeared to be a scattered collection of hobby farms. I knew the ranger and his family there and stopped by, as it was Sunday. The ranger's family had been members of one of the circuit pulpits my Dad had filled when I was going through 3rd to 6th grade. I wanted to say "Hi" and get some information, particularly on housing. I enjoyed some pie there with Rose, the ranger's wife, and learned that there was indeed a bunkhouse area above one of the large storage/work buildings (all painted white). As I was leaving, Rose said that going forward there would be no favoritism as the ranger wanted none of that or even the appearance of that. This was perfectly fine with me as I too did not want to be seen by my new work colleagues as favored or the ranger's "pet." At the same time, it was kind of gut wrenching, knowing that I was really out here on my own, sink or swim so to speak. I knew, and had told Dad, that if this effort in Montana failed, I would be returning to Stuart as a backup plan. Dad had strongly discouraged resorting casually to my "plan B." Ultimately, I knew, too, that in time of a crisis, the ranger or his wife would help me out.

Monday was chaotic at the ranger station with new and returning folk showing up for their first day of their summer jobs. There were mostly young men on timber marking crews, the interregional fire crew (hot shots), and the brush crew. My crew boss was a short, wiry Native American guy named Tom Boyd. Tom made sure to go over chainsaw maintenance, safety, and operation with me, as he must have known I was a bit green with chainsaws (plus I was an "out of state-er"). Maybe Tom knew I had some kind of connection to the ranger, but he never displayed that in any other form than making sure I knew what I was doing and was doing it relatively safely.

My chainsaw experience, which helped me land the job at 9 Mile, was cutting firewood for Dad's Franklin wood stove. Dad had a little McCulloch chainsaw with a manual chain oiler. This meant that you had to pump the chain oil plunger with your right thumb frequently as you cut. Dave Kiddel, a member of the church where Dad was the pastor, wanted us to thin out the black locust in one of his tree grooves. My younger brother, Craig, and I were to buck up all of the more substantial branches (greater than 1" diameter) into lengths less than 14 inches long and take the firewood back to our place in Dad's trailer. I think Craig and I spent 2 weeks or so on this project, but I did gain experience and skill with the chainsaw, even though it was a very small one with a short bar. No nicks or cuts from the chainsaw, but Craig and I kind of did sanity safety checks on whoever was running the saw. Importantly, we switched the sawing task whenever one of us of got tired. I think Dad had given Craig and me a safety briefing, as Dad (and Craig and I) saw many loggers in Dad's previous parish in Seeley Lake, Montana, with missing fingers. My right thumb, which pumped the chain oiler, locked up with muscle cramps one lunch while eating with Dave Kiddel. It was a bit startling, but we all just thought it was rather humorous.

The 9 Mile brush crew's main task was thinning younger trees so that the bigger or selected ones could grow bigger, faster. This also would have reduced fuel load in the event of a wildfire, but I am not sure that was a very high priority at that time—timber production was a higher priority. Sometimes the thinned trees were already felled (laying on the ground), and sometimes we had to cut the trees ourselves, leaving the paint-marked trees remaining. The timber marking crews paint-marked the trees for thinning and logging operations. We cut the downed wood into stackable sizes and then piled it into moderately sized mounds. We tried to locate our brush piles close to where the bucked-up wood lay and in more open areas where it would minimally impact the remaining trees when Forest Service staff burned the piles later in the winter. In the early winter the Forest Service staff would return and put a small piece of small black plastic over a section of the pile with some fine material that would burn easily. Then in the winter when there was snow on the ground, they would come in with drip torches and light the piles, primarily focusing on the fine, dry material protected by the black plastic.

A secondary brush crew responsibility was fighting small local forest fires. We were known as "ground pounders," not to be confused with the smoke jumpers based near Missoula. We spent time preparing fire packs (C rations, water, canvas half shelter, Pulaski—ax on one side, hoe on the other—, shovel, flashlight, batteries, etc.) and chainsaw packs (chainsaw, oil, gas). There was a local firefighting fitness test as well consisting of a distance run (in boots) and a certain number of chin-ups. We went to official fire training in the neighboring ranger district, Thompson Falls. After completing our largely classroom-based training, we piled into a large school bus and unloaded in the middle of an active fire. We filed out of the bus and were milling around as our leaders tried to get clarification on what we were to do and where we were to go. There was a loud crack and a large Douglas Fir started tipping directly toward our group and the bus. Folks, all in the yellow fire-resistant shirts and hard hats, were sprinting to get out of the way. Hard hats and other gear were tumbling, and no one was stopping to pick any of it up right then. One poor fellow tripped, fell down, and frantically half crawled and half ran as the large tree accelerated toward him. With wind rushing through the branches, the tree struck the ground only a foot or two behind the fallen fellow, narrowly missing the bus we had just disembarked from.

Toward the end of the summer, some of the hot shot crew members typically had to depart for their winter or fall jobs (some were school teachers). To complete the hot shot crew so it would be ready for a call to a big fire anywhere within the United States, selected members of the brush crew would be asked to temporarily join the hot shot crew. On my second or third summer of working on the brush crew, I temporarily joined the hot shot crew late in the season. I heard rumors that I was known as a hard worker, which improved my chance of getting on the hot shot crew. Near the end of my second fire season, the ranger told me that he was happy to have recommended that they hire me initially, as now everyone wanted me on their crew. That made me very proud, but as I recall, I just shrugged my shoulders.

There were multiple fires that I assisted with as part of the hot shot crew. One was north of Scottsdale, Arizona. On this fire air temperatures were HOT. One of our crew had to be hauled to the hospital with heat exhaustion and be packed in ice. On another fire out of McCall, Idaho, in a rough wilderness area, we had to hike out after a long shift as there was some issue with the helicopter that had dropped us off not being available to take us out. We had to cross a substantial fast flowing river at night (looked for a mono-rail that was supposed to be our crossing, but never found it). We hiked most of the night and ended up running our time for 42 hours straight. If they had offered me $200 per hour to work longer, I would have refused. I was absolutely spent and even felt ill.

Another extremely remote fire was in the Boundary Waters Wilderness in Minnesota. We flew in on a Beaver airplane on floats. There was a lot of deep peat, which was very dry and smoldering away with the most wretched smelling smoke. I got heartburn, constipation, and canker sores in my mouth near the end as the only food we were provided was canned C-rations. We were there for 2 or 3 weeks. Everyone on the crew was good and ready to head home. It was pretty there with the lakes and all, but the grub was hard to stomach long-term.

The most memorable fire was near Monterey, California. Our ride to our section of the fire was in an open army truck with bench seats along the outer sides of the truck box. We sat in single file facing the center of the truck with our backs to the wooded slats that formed the truck's box. As the truck wove through the narrow road, tree branches would fairly often warrant ducking to avoid injury. The guy in front on the bench nearest the cab of the truck would see the branch and duck, but the second and on down the line would all get whacked. I was constantly scanning ahead, curious of my surroundings and where we were going, so I avoided most of the tree branches, but others were not so lucky. Finally, after some guy got whacked the fifth or sixth time, he yelled in a very agitated voice that the guy in the front should shout a branch warning to rest of the guys on the bench. This obvious solution, however, did not occur until the second or third day.

The fire was high on a ridge and there was a small celestial observatory there. One guy, who seemed a little strange, decided he could get a rock all the way to ocean down below. He whipped out a sling, dropped a rock into the leather patch, and began whirling it over his head. Guys began running in all directions with hard hats falling. Everyone wanted to be as far away from this idiot as they could before he released the rock! Before most of us could get more than 10 feet away, he released one of the sling's strings aiming for the vast ocean down below in front of him. There was a loud BANG! Everyone stopped and looked around and to our amazement no one was hit. After a quick glance toward the ocean there was no rock airborne headed that way either. Scratching his head, the guy with the sling started looking around and found a dent in the observatory directly behind him. He was only off 180 degrees and seemed mildly amused, but everyone else was kind of pissed off and made him put his sling away.

This fire was in mostly chaparral shrub, which has lots of volatile oils that makes it extremely flammable. They had brought in loggers dressed in plaid flannel shirts, suspenders, and big white boots (a top of the line forester and logging boot is which is black in color). The local loggers had chainsaws to fell very large diameter trees. Their saws, which were to help clear the chaparral brush, had huge bars that had to be four to five feet long. It seemed ridiculous watching these guys cutting chaparral brush that had a basal diameter of less than 3-4 inches with one of these long bars. I only saw them once, on the first day, so I think the fire boss saw the folly in sub-contracting some of the work out to these local loggers.

The chaparral system had poison oak (or so we were told) in spots and we were warned to keep an eye open for it. About the second or third day on the fire, three or so of our crew were carted off to the local hospital. They apparently had inhaled smoke from burning poison oak. One or two of them came back to the crew after a couple of days, saying they were given a cortisone shot. But soon they were headed back to the hospital and I didn't see them again until we were back at 9 Mile.

Once we were hiking out along a fire line, headed to a new location where we were needed. We came across a spot where some burning material had rolled downhill and across the fire line. Our crew boss had us spread out and begin digging a fire line around the hot spot that was outside the fire line. There was a lot of fine fuel (mostly dried leaves on the ground) and chaparral shrubs in a small saddle where the hot spot was. I kept checking to be sure my emergency fire shelter (aluminum foil pup tent) was in my fanny pack and handy, as I was a bit concerned working in this volatile situation. My plan was going to be to kick as much of the fuel on the ground out of the way, deploy my fire shelter, and hope for the best. Suddenly there was the fire boss amongst our crew asking where our crew boss was. We pointed out our crew boss and the two talked briefly, then our crew boss yelled for us to move out. We headed single file out of the saddle and back to the fire line. I glanced back where we had just been digging a line around the hot spot and it was all engulfed in tall, roaring flames. We were lucky to be alive.

Another day we had been digging fire line all day and were tired and hungry. Our crew boss got orders for us to hike through the burned area up to an unburned patch on a ridge. It was after dark when we finally got there, only to find other crews all camping in fire slurry (pink, slimy fire retardant probably as harmless as agent orange?). The fire tanker planes had dropped several slurry loads on this location, and we were to eat and sleep in this pink goo. As I recall, the food was OK (not C rations, maybe some kind of freeze-dried stuff reconstituted), and they had paper sleeping bags for us. I was very tired and had little problem falling asleep, assuming the fire retardant must not be terribly toxic. We were hurriedly awakened in the middle of the night and ordered to evacuate as soon as possible as the fire was headed our way. We headed back through the burn area, which was still pretty hot and smoldering. We just had to keep moving or our boots would get too hot. It was a long dirty walk out, but no one complained. I never washed the slurry off my helmet, as I thought it was kind of a badge of honor.

There were some impressive views of this fire crowning out in the upper vegetation canopy down a moderate slope from the fire line we were patrolling. The fire line was substantial, a dozer line along the ridge top. Large, fire retardant planes dropped slurry close by (I assumed to knock the fire down as it approached us and the dozer line). It was very beautiful to see as it was near dusk and in full view from our position. I took some of my best fire pictures here.

I worked for the Forest Service through my undergraduate years. One of the summers I was with the Lolo Helitack crew, one year on trails (maintenance of back country trails, packing in with horses and mules, and helping with ground pounder fires), and one year doing grazing assessments in back country grazing allotments near Glacier National Park. I could make enough money in the summer to pay for most of my college fees when combined with university work study jobs during the school year, an impossible feat these days with high college costs.

— 3 —

FROZEN FEET (1974-1975)

Prescott College in Arizona was an outward bound "backpacking" type of school. Orientation was 30 days on some wilderness trip. My orientation was hiking with a group of 15 or so on the North Rim of the Grand Canyon, with multiple descents down to the Colorado River. During the last descent, along the banks of the Colorado, we all had to do 3 days solo and fasting. This was a tradition of several Native American tribes, and how they got their personal vision and their name. My "vision" came late in the evening of my second solo, fasting day. I could hear a helicopter (I was positive based on my previous summer working with the Forest Service fighting fires). I could see all of the sky around me, but no helicopter. It was getting louder and louder until it had to be right on top of me. I jumped up and there flying down the Colorado River just above the water was a helicopter full of tourists. Then everything started getting black on the periphery of my vision, narrowing down until I could only see the center of where my eyes focused. I decided I was fainting, so I crouched down with my head between my legs. I did not want to faint and bang my head on a rock! So that was my vision. No idea what it means, but I do know that after three days with no food, I was weak. I had been one of the stronger hikers in our group, but after that, not so much.

After the three-day solo, our group got back together and some pigged out on food and then got sick, while several of us started with small doses of mild food. Mine was oatmeal. Then I went swimming (well, wading) in the cold Colorado River. I was buck naked, as our group had done some co-ed birthday suit swimming back when we reached the Thunder River after a long stint with no water resupply. I was about ready to clamber out as I was getting cold, when I saw a McKenzie boat coming in with one paddler in a spray skirt. I stayed in the water, shy and unwilling to run to my clothes. Hopefully he would not stay long. Then another McKenzie boat and another, and soon there were 8 or so of them chatting it up on shore with my colleagues. I was really getting cold, especially since I was very low on fat reserves after 3 days of fasting. Finally, the visitors started piling back into their boats and headed down stream. I stumbled out of the cold water and got my clothes and coat on ASAP.

Prescott College courses were mostly pass/fail, and nearly every weekend they had backpack trips you could sign up for. Trips that I remember were Sycamore Canyon with old, small cliff dwelling like caves, and the Superstition Mountains where the lost Dutchman was rumored to have hid his gold. One frustrating trip was the ascent of Humphrey's Peak, Arizona's highest peak near Flagstaff. We were ascending the ski slope as the ground was bare at that time. As we got higher, but not quite to the top of the ski run, some other students started whining. It was too steep and there were snowflakes in the air. Some folks were cold. The leader of the group asked for my and one of my friend's recommendation. I said let the whiners head back to the bus and let's proceed with only the better prepared hikers. Apparently, we all had to stick together (safety first!), so we started descending. As we were descending, I was pissed! I was walking to the back of the descending group, next to the expedition leader (same guy each weekend so he knew me fairly well). I pointed out to him that all the tired or cold whiners were laughing, trotting, and playing on the descent. This really made me and my buddy mad. We decided we would do our own assault of the Peak.

Back on campus I checked out a winter sleeping bag. I had cold weather gear that I had confidence in, except for my hiking boots, which Dad gave to me because they were too cold for him. My buddy told me about vapor booties. You put a plastic bag over your foot, then your socks over that, and finally another plastic bag. This sounded like a good approach to me. I had snowshoes checked out from the college, and my buddy, Wayne, had crampons. We bought a Cornish game hen and a can of generic beer for a Thanksgiving Day meal, which would occur when we were in the back country of the San Francisco Peaks region.

We got to the bottom of the ski run late in the evening, it must have been a Friday night. Snow was on the ground. We trudged up the ski hill and nightfall soon came. Finally, the slope flattened out for a short distance, possibly an old road to access the ski lift in the summer. We pitched our tent and were late getting up the next morning. It was cold and we thought we would let it warm up a bit. Then we started hearing swishing sounds outside the tent. I snuck a peek out the door to see skiers swooshing by. We broke camp.

As I recall breakfast was not much (granola bars and water?), and we were soon panting our way up the ski slope again, but this time with oncoming skiers. You may have heard how sometimes runners become agitated and aggressive with cars on the road? Some kind of endorphin effect from extreme exercise? Anyway, I think I got that going up that last stretch of ski hill. I was prepared to hit those skiers like I would make a football tackle, with all my strength if any got too close. In retrospect, we were invading the ski hill and should have been off on the edge or even off the ski run in the trees (which would have been MUCH more difficult for us).

Somehow, we made it past the top of the ski run alive, and the summit appeared to be rather close. Wayne had his crampons on, which gave him a little advantage on the ski run. On the ridge line up to the summit, it was hard, wind-blown snow that I had problems walking on. I fell more than several times, and on multiple occasions started sliding and accelerating downhill as soon as I fell. Fortunately, Wayne had made sure we both had ice axes, and I had to execute a self-arrest several times to keep from accelerating on down the hill. It made for anxious climbing, which continued up to the summit and as we began the descent down the north side. I was in front and Wayne said, "Hey, where's my tent?" I had insisted on carrying Wayne's fancy tent as I thought it was macho to have the tent up on top of the pack frame. I then recalled that on one of those self-arrests, I had heard a "zzzzzip" and looked around when I got up, but saw nothing. The tent must have fallen off then. I dropped my pack, grabbed my ice ax (I was not going anywhere without that!), and pretty much sprinted back up to the summit. Wayne's tent was a top-of-the-line four season tent!! I thought the "zzzzzip" incidence occurred around the summit. I went down the side slopes aways, hoping to see the tent against some rock, but no cigar! I returned to Wayne with the bad news. Wayne did have a black plastic ground cloth, so our plan was to do a glissade down to timber line on the north side of the peak, sliding on our butts and using our ice axes for steering and breaking. Once down near timber line and all in one piece, we laid out the ground cloth and our sleeping bags, had some supper, and went to bed. I was exhausted and my feet were cold. I decided to keep the vapor booties on overnight, and just remove my boots. It was a COLD night, and the down in my sleeping bag seemed to just puff up around the upper edges of the bag, leaving very little between the cold ground cloth and me and my feet. My feet were cold, and I was having a sleepless night. Finally, I just resigned myself to sleeping with cold feet. I felt really bad about losing Wayne's tent as well.

    Next morning, I awoke but found that I could not feel my feet. They felt like pieces of wood. I took my socks off and my vapor booties to see what was happening with my feet. To my horror there was ice which froze the bag to my feet. The sweat from my feet from the hike up had frozen to my foot. I warmed my feet by the fire, and several times Wayne had to warn me that my feet were too close to the fire. I do not remember if they were discolored at this time, but they would turn blue on the entire length of all my toes before we got to the car.

    We hiked three days, kind of traversing around the base of the mountain and over a saddle going northwest. With the snowshoes, I had little problems proceeding in the soft, hip-deep snow while Wayne floundered in the deep snow. When we did come out at the ski lodge, we went inside to discuss what to do. The ski lodge guy wanted to call an ambulance, but I was sure that would be expensive, and I had just walked 3 days on my feet to get out, so I could not understand why we could not drive the hour or more into Flagstaff ourselves.

At the hospital, what they did was de-breed all the blister that had formed. They removed the old dead skin and drained the blisters that still had fluid. They must have given me some antibiotics and pain medicine, but my memory of that is vague. Back at Prescott College, there was a college speaker or movie that my friends took me to (carrying me using the fireman carry). Upon return I found a wheelchair that my aunt and uncle from Socorro, Arizona, had dropped off while I was out. I had called Mom and Dad and they must have called my uncle and aunt. It had no foot rest, as it used to belong to an amputee. I put a strap across the lower front frame of the wheel chair where I could at least hook my heels. I would ride around campus, which had all these little flights of stairs (two or three here, five or six there). My approach was to get as close to the stairs with the wheelchair as was safe, and then stand up and move it down the stairs and then get back in. I could feel nothing, and the doctor had told me I should be in bed, off my feet.

There was a class at the college where they were instructed to do something abnormal for a certain number of days and see how other students reacted. One of my friends knew a girl who was not speaking as part of that class exercise. I could not have cared less, but one day while moving my wheelchair down the stairs some student who I did not know accused me of faking my injury as part of that class exercise. He let me know that he thought either I was being deceptive or trying to draw attention to myself and marched off. I decided I didn't care what he thought, but hoped someone let him know later that he had put his foot in his mouth big time.

We had an "all hands on deck" college meeting at the auditorium one night. I was not going to go. Too many steps and I was supposed to be bedridden. But my friends came and got me, doing the fireman carry again and placing me in the back row. The college president got up and informed us they had gone bankrupt and that all students had to be off campus in a day (or two?). They were paying students that worked (I was work study doing dishes at the school cafeteria for breakfast) with cashier's checks, so we could cash them after the news about the college closing broke. I got my cashier's check that paid most of what I was due and would be just enough for gas to get me home to Stuart. I am not sure how I moved all of my stuff to my car. Maybe my friends helped or I was able to get illegally very close to my room with my VW bug (manual transmission). So late one evening I headed out, cashier's check in hand, and a nearly empty gas tank. In the second small town I came to, I stopped at a bar and asked if they would cash the check. The bartender was aware of the college closing and seemed to think I was trying to trick him out of money in the till. I pointed out that the college intentionally did cashier's checks so students could get home. I told him my distant destination and empty pockets story. Things were looking bleak, and I was pretty sure I had a snowball's chance in hell of getting my money here. Where else to go or do? How far before I just ran out of gas? I started tearing up and my voice breaking up. To my complete shock, the bartender cashed my check! I thanked him profusely, filled up my gas tank, and was on my way. It was winter and thankfully, no really bad road conditions. I did not eat as I wanted to be sure I had enough gas to get home.

When I got home, Dad focused me on attending another college, one my younger brother Craig liked, Yankton College in South Dakota. Dad did not want me laying around just healing, so I went up to Yankton, to look at the college. I was still supposed to be in a wheelchair, but I was again cheating and actually on crutches. After a quick tour, I was accepted and they transferred as many of the Prescott College credits (one whole semester!) as they could. I attended Yankton one semester, but later decided to go to the University of Montana at Missoula since I was primarily interested in wildlife management.

It took six months for the skin on my toes to grow back. When I started back fire fighting for the Forest Service in the summer, the toes were not quite healed, but I told no one. My feet never really hurt. I must have frozen them just right, to deaden the nerves. But the hang nails, when all my toe nails began growing back, were PAINFUL. The doctor removing the hang nails dug like crazy into my toes, but missed a few, so I have some weird slivers of toe nail which grow 90 degrees off the toe. As a result, I know right away when I need to trim my toe nails because putting on or taking off socks is painful otherwise. No apparent cold weather effects seem to persist, as I continue pursuing winter camping, and downhill and cross-country skiing.

— 4 —

ANDY AND AFRICA (1975-1985)

After deciding to drop anchor in the Missoula area for both school and summer work with the Forest Service, I looked for long-term housing. Initially, I jumped around a bit, but ultimately the fire lookout guy at William's Peak, Craig, was moving out of his Missoula apartment and heading back home. He suggested I check out his rental place, as it was only two blocks from campus and rent was only $30 per month with utilities included. There were two things he cautioned me about: 1) unfinished basement ceiling for the kitchen and common area, and 2) no women allowed. The place looked fine to me, but I was a bit unsure of Andy, the older fellow renting the other bedroom and who I would be sharing a refrigerator, kitchen sink, and a common bathroom with. We each had separate kitchen tables and electric hot plates. But as time passed Andy and I became pretty good friends. I believe I stayed in that basement for four years, below the land lady's single-story brick house.

Andy worked at a local department store's gun department, and he was pursuing a range management undergraduate degree. I was initially fixated on a wildlife biology degree (driven in part by several books I read in high school by Farley Mowat, *Never Cry Wolf* and *Owls in the Family*), but Andy made a strong case for a range management degree. He pointed out the very competitive job market for wildlife biologists, the way a range management degree could indirectly (through habitat) link to wildlife management, and the higher probability of a job post-graduation with the range degree. So, I soon changed majors to range management, being also impressed by ranchers Dad knew both in Nebraska and Montana.

My impression was that Andy also was a fairly good writer. I would have him edit my research papers I had for various classes. He took the time to edit and make suggestions on some pretty rough drafts, but slowly (and reluctantly) my writing improved. I distinctly remember having a major required paper for a soils class later in my studies. When Andy reviewed it, he took it into his room and came boiling out within 30 minutes and slammed the paper down on my kitchen table and said, "You know how to write!" and went back in his room and closed the door. I got an A+ on that paper. The Professor announced to the class, after handing back our graded papers that one of us had improved exponentially from a previous paper. He was asking for some kind of signal so he could use that student as a stellar paper example and an example of improvement. I knew he was talking about me, but I remained motionless and expressionless. He ultimately had to use another example.

Andy had been a Peace Corps volunteer in Botswana, Africa. During my final year at the University of Montana, he suggested I apply for the Peace Corps. I thought that I had done pretty well going from a newbie to getting established with a network of friends in the Missoula area. So, I went ahead and sent in an application. Dad and Mom were both quite excited. My grandfather on my father's side had been a missionary in India, so I suspect that and his persistent curiosity drove Dad's strong support and enthusiasm.

The Peace Corps process was to send you three different jobs and you had to select one. If you did not like any of the three, you were done. The three job positions I got were helping vegetable farmers in the Caribbean, a forester in the Congo, and another forester, I think in Cameroon. I wanted something where I could apply the range management skill and knowledge I had just acquired, so I declined all three. I thought that was that. Shortly after that, I got another single job announcement from the Peace Corps for a range manager in Niger, Africa, working with pastoral tribes in the Sahel. This looked very promising, and I had no job offers in the States and few of my fellow graduating class members were finding jobs. I mentioned this Peace Corps job to the ranger back at 9 Mile and he said, "They could drop you anywhere in the world and you would do good." With that strong vote of confidence, I accepted the Peace Corps position. It turned out that this particular Peace Corps job was keenly focused on range management, and they were having a hard time finding folk with that specific academic training, so I was also exactly what they were looking for.

There was a flurry of mail exchanges about paperwork I needed to send in to get a passport and Niger country approvals. There was also a required range management training session in Mancos, Colorado. To meet this training date, after my last final exam, I had to be on a plane to Manco within 12 hours. So, I had to store a bunch of stuff at Mom and Dad's, who now were at a parish at Wise River, Montana. With them being so much closer, it was easier for me to leave my stuff with them. It was kind of a blur with finals week and blasting off ultimately for Africa so quickly. I later learned that this was intentional, as the Peace Corps did not want me to find a stateside job after graduation and change my mind.

After the range management training at Mancos, where I learned a lot and also impressed them with what I knew, it was on to Philadelphia for more general Peace Corps training (cross culture and some insights to what language training would be like) and vaccinations. I was really out of my comfort zone in Philadelphia. I was a country bumpkin in a big city. It was then on to Paris and then a flight down to Niamey, the capital of Niger.

I had a tough time learning one local language and French in one month. Of the 60 trainees, I was second to the last in language skill after several weeks of training. They were worried I might get frustrated and terminate early and return to the States. One of the guys at the stateside range management training was helping with the new Peace Corps training in Niger and visited with me about how I was doing and how badly they wanted me. I assured him that I was in this for the long haul and would somehow get over this language hurdle, maybe not in a month, but a bit longer. I ended up staying in Niger for a cumulative time of 10 years, which I suspect was much longer than any of my fellow other trainees.

Andy had a rather significant impact on me and my trajectory through time. Many years later, I was at a Range Society annual meeting in Rapid City, South Dakota. Andy was listed in the program as a presenter in one of the sessions. I attended his presentation on oil pads in National Grasslands in North Dakota. I stood in the back of the room and when Andy started down the center aisle to exit after his speech, I intercepted him midway, intentionally causing him to turn to avoid running into me. I turned again right toward him, blocking him again. He stopped and looked at me, more than a little pissed. Extending my hand and said. "I am Bruce Wylie." He blinked and said a very surprised, "Yes you are!" We went out in the hall to have chat for 20 minutes or so.

— 5 —

ADAMOU (1979-1980)

I survived Peace Corps in-country training, BARELY! It was stressful, and I really had a hard time with the one language, much less two (one local language, Housa, and the official language of Niger, French). I did well in neither. I remembered, reluctantly, my earlier certainty in high school that I would never leave the U.S. Clairvoyance appears to be an attribute rarely bestowed on high school seniors. At the end of Peace Corps training, I went to the project office that I worked for, which was in Maradi, the same city where the in-country Peace Corps training was held. I was planning on introducing myself to the project director, who had attended college in California. I assumed I would again deal with people who I could not communicate with. As I walked by, several Nigerien men were standing around the front door of the office. As I walked toward the front door, someone asked me how I liked Niger, or something similar, in perfect English. I scanned the cast of characters at the front door and only one of them was smiling with friendly eyes. I stopped and I responded with something intended to test his English (many only knew basic greetings in English, similar to my Niger language skills). Whatever my test response was, I was floored by the same accent-free, logical English response. He had my full attention. We chatted briefly, and then I went in to meet the project director.

Dr. Ali was sincere and welcoming and clearly summarized the project goals. He said that my first assignment would be to go with Bud Rice, an

American, up north, where he would introduce me to the "project zone," or our study area. I do not remember my fellow Peace Corps volunteer partner Karen, who I first met at the technical range management training in Mancos, Colorado, being along. My recollection was being alone and feeling a bit hung out to dry solo on my first visit to the project office. Dr. Ali escorted me out to find Bud Rice. As we exited the office, the English speaker fellow joined us, stating in English where we could find Bud, out by the vehicles on the project campus. The project campus was crammed full of new white Scouts and Scout pickups. These had really fat tires (I assumed for driving in sand) and heavy (almost armored) skid plates under the engine compartment with an oversized gas tank under the pickup box in the back. I did not count them, but there were 30-40 all crammed within the project office concession walls. We found Bud and he told me of our departure date for the north country and what I should bring. Bud had been a Peace Corps volunteer in southeastern Niger and spoke two local dialects. A Nigerien named Adamou had worked for him as a house boy. I listened closely, as he seemed to know what he was talking about.

Adamou and I rode in the back of Bud's pickup loaded with gear (water bottles, gas jerry cans, camping supplies, personal gear). Karen was in the cab with Bud. Adamou told me he could teach me Housa (even though he was a Djerma from southwestern Niger). I was interested and asked how. Adamou said I needed a small pocket notebook and a pencil ALWAYS in my possession so I could write down new words to add to my word list for the day. Each day I was to memorize five words. I was to re-study the words many times per day to ensure I would have them well memorized by the end of the day. Adamou said I should write the words however would best help me pronounce them correctly. After two days of this, as we drove north out of civilization away from town into the bush on two tracked sandy roads, Adamou said, based on my memorization success, I should now do 10 words a day.

We were heading north from Maradi to a small village called Godabedji (where the previous Peace Corps volunteer, Warren, had lived and rumor was, where I was to be posted). We visited the Toureg chief there, Kinni. It was fairly formal and I tried to be on good behavior. After spending a day or two at Godabedji, we headed north to the even smaller town (?) of Tchin-Ta-Borack where we turned east, headed to Aderbissinat, camping somewhere along the two-track road. The next day we arrived in Aderbissinat and met another Toureg chief named Boha. Boha was 40 to 50 years old, similar in age to chief Kinni. Later, Boha married Kinni's daughter who, by guessing, I would say was high school to middle school aged.

About the second day, we camped at Boha's camp just outside Aderbissinat and headed into town. It was market day (one day a week a town would have its market day, the same day of week, every week). As I was getting out of the Scout pickup box, Adamou said, "Here is your language test. Go buy some onions." I complained that I did not know the word for onion, but Adamou said, "Ask and figure it out." Off I went, thinking I could just look for onions, but quickly realized I would be here all day visiting every aisle in the market. So, I stopped where a guy had dried tomatoes laid out on a blanket. I asked him where the thing is that you use to make sauce, but when you cut it, it makes you cry. Three or four guys were laughing and saying that they had no idea what I was talking about. But the guy selling the tomatoes (about my age) stood up and yelled, "Albasa!" Everyone else was quiet and stunned. I had no idea what albasa was. I asked if he could take me there. He led me off to the onion seller. I thanked him profusely and he just smiled.

I returned to the truck, proud and confident, with a bag of onions! My vocabulary was reaching the level where I could ask what a word meant. Now, I could build my own daily list of 10 words to memorize! With vocabulary, I could communicate! I had a vague understanding of tenses and sentence structure in Hausa from the Peace Corps training, it was just a deficit in vocabulary that was my major problem. I could live with breaking the grammar rules! I was concentrating on vocabulary. I could reinforce grammar improvement by listening to others.

I learned many things from Bud and Adamou: fence building (how to make 90-degree fence corners using the Pythagorean Theorem—Bud was a math teacher—and braced corner posts from metal posts), customs, taboos, and just to relax and treat people as people. Just forget the folks that are not patient with my accent and weak vocabulary. There were plenty of people that would be sincerely interested in helping me learn their language.

— 6 —

TRADITIONAL WELL (1979-1981)

I was traveling with Bud and Adamou on a two-track jeep trail which led north of the town of Dakoro in south-central Niger to Aderbissinat, which was southeast of Agadez. Boha was a Toureg chief near Aderbissinat. We were preparing to build some grazing enclosures in the area and wanted to be sure the chief and the locals were OK with it. The enclosures were meant to see how the vegetation would respond to no grazing or browsing pressure. Bud liked to stop at random locations to spend the night with apparently random pastoralist camps. I kind of felt like we were intruding as we were not invited. But the camps seemed flattered to have us as guests and would often slaughter a goat or sheep to celebrate our visit.

This particular time we were at a Wadabe (Fulani) camp and it was early afternoon. I partook in the traditional three sets of tea (first set very strong and the third set mostly sugar water and all sets of tea were drunk from shot glasses). Conversation seemed to be waning, and I was more curious about what folks did in their normal daily activities, not so much the pomp and circumstance of tea. The man of the camp had told me where the well was, and this was where the young men were watering their family herd of cows. The well usually was just a hole in the ground, 5 to 6 feet across. The depth of wells varied depending on how deep the water table was. This particular well was about 25 feet deep. Animal traction was used to pull the rather large leather sack-like bucket up when it was full of water. Typically, a large steer, a camel, or several donkeys tied together did the pulling.

On this occasion, 3 donkeys lashed together were being used to do the pulling. Around the top circumference of the well would be two to four substantial forked logs or branches angling a bit toward and over the edge of the well. The "Y" end hanging over the mouth of the well had small (3/4"?) holes bored through the wood near the ends of the "Y." A wooden axle was inserted into these holes with a wooden pulley (~8" diameter). The rope from the pulling animals went over the pulley and down the well, with the large leather "sack" at the end. When the leather sack full of water came up, the pulling animals were ordered to stop (often a younger kid led the pulling animals to ensure they did stop). Young men (usually two) would grab the leather sack, untie it from the rope, and carry it over to a watering bowl. Once the water was poured, the young men would yell and bang the horns of the cattle, such that only one drank at a time, not spilling the water. One major problem with this was that the rope dragged on the ground as the empty leather sack was reattached to the rope and thrown into the well. The traction animals walked back toward the well and got ready for the next load of water. Recall that there are herds of cows, goats, sheep, and camels all standing around waiting to drink. So, there was a lot of animal manure. The dry dirt and manure would stick to the wet rope as it slithered across the ground, ultimately contaminating the well.

It was hotter than blazes, as we were into the hot season now. The grass had all seeded out and mostly was a dried brownish yellow color. I had cheap rubber flip flops on so I had to watch closely to avoid stepping in cow shit or on tree thorns (some of which were over 3 inches long). I was walking parallel to the rope as the three donkeys pulled with all their might. The 2 muscular young guys at the mouth of the well were staring at me. I was kind of getting used to being stared at being an American and white in a strange land.

There was a loud "POP" and everyone started yelling and they were either very mad or very alarmed. The rope slithering next to me was accelerating and the wooden pulley at the well was whirring loudly. I thought

I could help by stepping on the rope to slow or stop it before the leather bucket hit the water at the bottom of the well. After all, I wanted to help. But something did not seem right. No one else was running to grab the rope and all just stood clear and watched. I decided to do nothing and watch and learn. Wham! The bucket hit the water in the bottom of the well. The guys at the well started yelling at each other and one went out to the pile of rope and came back with a grappling hook. They tied that on the broken end of some rope and fished around with it at the bottom of the well. After snagging the leather bucket and pulling it back up, they tied the grappling hook rope to the donkeys and they pulled up the half full leather bucket which was tipped off to one side. They made the necessary repairs to the rope and to the attachment on the top of the leather buck. In no time, they were back in action, watering the rest of the thirsty herds.

I did not think much of it but when I returned to the camp told Adamou what had happed. I was mostly just impressed how they solved their problem. Adamou, however, said I was lucky to still be alive. He said many people get killed in that situation. If that rope rapidly receding to the well, and pulled by a large mass of heavy water, wraps around your foot or leg, you too are headed over the Y wood frame and pulley and then down the well. Another near miss with death. I felt stupid and was never going to let myself get in that predicament again. I suspect the young guys at the well were watching me because they knew the rope had a weak section and they knew I was in the danger zone. They were busy and did not want the trouble of trying to communicate with an ignorant foreigner.

Some of these open wells were cement lined. They have a circular form where they pour the cement in and let it set up. Then they get inside the circumference of the cement ring and begin digging out the middle and under the cement ring. The cement ring slowly descends as the dirt is removed until the top of the cement ring is near the ground's surface. They then set up the cement mold on top of the existing cement ring in the ground and pour more cement in the mold. After it dries, the digging begins again.

At the small village of Godabedji, they were building one of these cement wells for the village near the school. I was again curiously checking out what they were doing and they asked if I wanted to ride the winch down to the bottom where two guys were digging away. Sure, I was game—some people never learn! It must be cool down there, I thought. As they lowered me, my curiosity captivated me. I admired how far they had dug and wondered how much further before they found water. But as soon as I got to the bottom, the bucket I had just rode down on was loaded with dirt and retracted back to the surface. With a heavy load of dirt on a hand winch above me, I suddenly became claustrophobic. It was hot and muggy down in the bottom of the well. No fun. I'm sure I was in the way of the two guys who were digging. I was more of a hindrance than help. I think we were 20 to 30 feet below ground. I let the fellows up on the winch know that I wanted out and they asked for confirmation. "Yes, yes, get me out of here!" They patiently lowered the bucket, smiling all the while at the silly foreigner. They slowly pulled me up. I was happy to get back to the surface and thanked them for letting me go down. They seemed to be kind of smirking a bit, trying to hide it as best they could. No doubt they had seen this drill before.

Later, when I told my story to someone in Niger, they asked if I could see the stars from the bottom of the well. I had never looked, but found out much later this is not even true!

Once I decided to ride my camel with one of our herders, Moukhamadoune, to the Catholic Mission, 14 km (?) or so north of the government ranch where our grazing trials were. Our herder was going there to get a large gunny sack full of millet (maybe 70 lbs. or more?). It was the hot season and I took my two-liter water bottle along, knowing I could get water from the French nuns at the mission. My water bottle was an old two-liter motor oil plastic jug. To prevent or mitigate drinking scalding hot water, you could buy these water bottles in the market with a couple layers of burlap nicely sown around the plastic jug. You only had to dampen that burlap and it would cool the water in the jug nicely.

Moukhamadoune paid for his bag of millet (I suspect at a bargain price), but I tried not to be nosey about the payment. We tied the big bag of millet behind the hump of Moukhamadoune's camel. He then sat on top of the bag of millet all the way home. My camel had a saddle in front of the camel's hump. On the way home, that saddle felt like a toilet seat that just kept rubbing me all the way around the rim. It was HOT on the way home in the afternoon and we ran out of water. Moukhamadoune knew of a well on the way, but it was a Fulani well. He was Toureg and doubted he could get water there, but he said I probably could. Sure enough, the Fulani filled my water bottle. Remember how these wells are contaminated by the rope dragging the dirt, animal pee, and fecal matter on its way back down the well? I did not care, the water was cool, muddy, gritty, and GREATLY appreciated. Once we were out of sight, I gave Moukhamadoune my water bottle and he drank. When we finally got back to the second well, I was shot, but tried not to show it. The next day? I was sore everywhere, especially all around the rim on my hind end. In confidence, several days later, Moukhamadoune fessed up that he, too, could barely get out of bed for a day or two afterwards.

These traditional wells are crucial for livestock and people out in the bush. In the rainy season (June–August) there are lots of seasonal ponds that provide sources of water both for people and livestock. The early dry season, the cool season (November–January), has cooler temperatures and harmattan winds that bring persistent dust storms. I learned later that satellites were able to track these dust storms across the Atlantic Ocean to the Caribbean Islands. In the cool season, shallow wells (6 to 10 feet deep) were dug in the dried up seasonal ponds. The deeper traditional wells are used later in the hot season (March to May), as the labor demands are significant.

— 7 —

BUCKO THE CAMEL (1980-1981)

The Peace Corps had a program where they would buy a camel or horse for volunteers who needed transportation for their work. I had a myriad of potential applications for a camel dreamed up. So, I crafted a handwritten letter to the Peace Corps asking for funding to buy a camel. I had a full page of application and needs for the camel: 1) Transportation back-up if our vehicle failed, as there was no busses or taxis that passed by. We were in the middle of nowhere and with no village; 2) Assistance with moving our cattle herds. In the beginning of the rainy and dry seasons, we would get new heifers and milk cows for our grazing trials, but often this occurred at the headquarters well and we had to move our newly formed herd (animals unfamiliar with each other) back to the second well. Some of these young heifers were headstrong and athletic and would try and run off; 3) Diet assessment to better understand desired tree and shrub species browsed on by camels. I had at least 8 or 10 reasons why I needed a camel.

After three weeks, I got a letter from the Peace Corps rejecting my camel proposal, saying that we had a vehicle at our disposal, but they enjoyed my colorful proposal. This irked me, but in hindsight there were a lot of volunteers who did not have a vehicle, so we were lucky. Being stubborn, I just decided to buy my own camel. I got approval from the ranch director and then headed to In-Tuilla on market day to purchase a camel. I took two of our herders who were Toureg (supposedly camel specialists, while the Fulani were supposedly cattle experts). They told me to stay behind and shop in the non-livestock section of the market, as my presence would drive up the price of any camel I or they were trying to buy. They found a healthy middle-aged camel with a fair disposition and negotiated a price. They came and got me and went over and paid something like $80 or so and pocketed the receipt for the camel. I bought an un-ornate working camel saddle and then one of the Toureg herders rode the camel home to the second well.

After letting the camel rest overnight, the herders brought the camel by and I had my first ride. No big deal, as I was used to breaking and training horses since I was in 5th grade. But there were differences between a horse and a camel, I learned. The camel saddle sits in front of the camel's hump and the rider's feet brace the back of the camel's neck. The rider's feet bracing against the back of the camel's neck is the only thing that keeps the camel saddle from sliding forward, where the camel's torso diameter is smaller. Without the braced rider's feet pushing the saddle back, the saddle would gradually move forward a little at first, but then faster as it moved forward on the camel, until the rider is ejected unceremoniously to the ground (camels are tall). If the rider's feet slip off the camel's neck, he or she needs to quickly re-secure their feet on the camel's neck right away. For this reason, many camel riders ride barefoot with their big toe pinching the camel's neck, thus helping to prevent the rider's feet from sliding or bouncing off the camel's neck.

Bucko, my house, and me.

Dana, one of two new Peace Corps Volunteers who joined me at the Dakoro Ranch suggested the name "Bucko" for the camel and I agreed. I asked Moukhamadoune to give us—Jay, Dana, and me—a demonstration ride to show us what Bucko could do. He was on board the saddle in the blink of an eye. Bucko was up and off at a reluctant gallop and he groaned in complaint, which only brought a snap with the whip. Moukhamadoune had Buck running around my hut and around trees at a full gallop. A camel is very tall and skinny compared to a horse. This necessitates a dramatic lean by a camel when cornering at a full gallop.

Later, we did a bite count study by following Bucko around recording how many bites he took of which shrub, tree, or grass species. We also used Bucko as a cutting horse. To be perfectly clear, the difference between riding a classical cutting horse, a Quarter Horse, and Bucko was similar to the difference between driving a sports car and driving a heavily loaded semi-tractor trailer. Nonetheless, when we moved the newly selected test heifers from the ranch headquarters to our grazing trial pastures at the second well, Bucko was invaluable. Invariably there would be one or two strong-headed heifers who would break from our herd and head off at a run to where their former herd was. Moukhamadoune was the ultimate camel jockey and pursued each and every breakaway heifer on Bucko. We never lost a single heifer in our cattle drives to and from the second well.

As my Peace Corps tour was coming to an end, I asked Moukhamadoune, to sell Bucko at the In-Tuilla market. Bucko had stepped on a thorn or piece of wood that we had tried to remove, but we must have missed some. Consequently, Bucko had a noticeable limp, favoring one front leg. The only bidder on the sale of Bucko was the butcher. The sale amount was less than what I bought Bucko for, but he was lame. I was sad to lose Bucko, but I had a long list of things to tie up before I headed home to the U.S. I will say that with the sale of Bucko I also experienced an immense feeling of relief, as if a ton of bricks was removed from my shoulders. The range consultant from Colorado had told us that when you have livestock you are always thinking about them—where to find feed and water for them. Also, I had been concerned about Bucko getting sick, injured, or even stolen.

— 8 —

MY FAMILY VISITS (1980)

After about a year in the Peace Corps, Mom and Dad wanted to come visit. Dad had let it be known to me that he was very proud of my being in the Peace Corps. He mentioned something about his buttons down the front of his shirt popping off as his chest swelled with pride when he told others of my Peace Corps experience. I had bought a bunch of photo printing envelopes from a company in the U.S. before I left. I would send the film canisters in the mail and have both prints and slides made, both going to Mom and Dad's address. I had asked Mom and Dad to send me the prints in Africa and to store the slides, that way I could give people the pictures I took of them. This opened up the doors for allowing me to get various pictures of some very casual situations. As a result, Mom and Dad were seeing every photo I took and enjoying it. Occasionally, Mom and Dad would write, asking for the story behind a particular picture. In short, I was unknowingly providing lots of "bait" to get them to come visit Niger.

Dad's interest in my Peace Corps experience also probably stemmed from the fact that his mother and father had been in India for my father's early years. My grandfather on Dad's side was a missionary and my grandmother was a nurse. Dad's older sisters remembered a lot of experiences and even some Hindustani words, but Dad was the next to youngest and only remembered a couple of incidents from India.

They asked what time of year they should come and I recalled that June was the start of the rainy season, a time when the grass would be green and the ponds full. However, perhaps I was recalling the rainy season in the southern capital city of Niamey where the rains come earlier. Anyway, my parents were going to be arriving in the late dry season when all the grass was gone and the livestock skinny as a rail. I learned that both my older brother Allan and my younger sister Darlene were also coming. I had let my Nigerien project chief in the southern town of Maradi; the ranch director at the Dakoro Ranch; the USAID overseer of the project, Paul Daily; and the Peace Corps know my family were coming. As the date of their arrival in Niamey approached, I headed down to Niamey to pick them up.

My family's flight into Niamey was delayed a day, which gave me another day of vacation in the big city. I do remember seeing my family from behind the movable fence-like barricade in the airport, which prevented locals from getting into the customs and passport approval area. I could see my family and was excited. They were smiling and waving. Dad must have been one of the first of them through the passport approval area, which was only 9 feet or so from the fence barrier I was behind. They knew no French, and I could see Dad was having an issue as the official kept asking him questions and he had a confused look on his face. Dad, while talking to the official, turned and pointed to me smiling. The official motioned for me to come over. I worked my way around to the gap in the barrier fence where the passengers would exit and walked toward the official who motioned me to come. Another policeman started yelling stop and told me that I had to wait behind the fence. I returned to my previous position on the fence close to where Dad was still talking with the official. Again, the official motioned to me to approach, so this time I just pulled the fence partitions apart enough so I could step through. Again, the policeman yelled at me and had me get back behind the fence immediately. I complied. The official dealing with Dad finally walked over to me and asked if Dad was my father and what was his occupation. I told him Dad was a preacher in a church. The official returned to Dad, and soon Dad was allowed to proceed beyond the barrier fence. None of the others in my family had any problem getting approved to enter Niger. Later we found out that Dad had put "Minister" as his occupation. In Niger, a minister was a high-ranking politician maybe akin to a head of a government department or agency. My mother said she thought that the policeman was going to shoot me for going past the barrier fence. Alas, this was only the beginning of the drama that would ensue with my family's visit to Niger.

I do not remember what kind of transportation we had from Niamey to Maradi, but I suspect it was one of the project vehicles that would be heading that way. Anyway, we had several days in Niamey before we headed to Maradi, where the project headquarters was and where we could get transport to the Dakoro Ranch where I worked. We stayed at the project guest house in Niamey, which was an apartment above a bakery. In Niamey, we visited the Grande Marche, or Big Market, which was a maze of narrow paths between mostly tin shack boutiques with vendors selling about everything under the sun. We usually had a contingent of local kids shadowing us and begging for money. I had warned everyone about pick-pockets, so everyone was "aware." I was leading this long line of white foreigners who spoke only English through the market, which meant we were very obvious and there was next to no possibility of getting a fair price as prices were negotiated on a case-by-case basis.

I wove my way through the market, trying to expose my family to the diversity and the style of an open-air market. I do not think my family bought much as they were trying to travel light. After a couple of hours, I was getting a bit tired and thirsty and stopped to see if everyone wanted to go to the apartment or go to a restaurant. To my, and everyone's horror, Dad was missing! Dad has a curious mind (like me?) and likes to talk to folk, so maybe it was poor planning having him at the rear of the entourage. But it's odd logic to think that my brother, Allan, should have provided adult supervision for Dad. I told everyone to stay in place as I tried to retrace my tracks a bit looking up and down every passage hoping to catch a glimpse of Dad. No luck. I tried asking some of the kids and merchants if they had seen an old, white-haired white guy. No one had. At a loss, I returned to the rest of the family.

I thought out loud, laying out several options. It seemed that continuing to wander around the market maze had a very low probability of success. In the end, we decided to return to the apartment and wait. I was thinking that I might try and return and search for Dad if he did not return by the end of the afternoon siesta. Mom, remarkably to me, did not seem concerned, claiming Dad could find his way back to the apartment. I was extremely doubtful. We had only been at the apartment for 30 minutes or so when Dad walked through the door of the apartment! Flabbergasted, I queried how he found his way back to the apartment with only English and only a scant exposure to Niamey's urban geography. When Dad realized he was lost and that he could not find us, he looked at the city skyline and saw one of the few tall buildings (a short skyscraper?) that was near the apartment. He navigated toward that and then saw some landmarks he remembered, which guided him to the apartment.

Dad's "adventure" contrasts with another episode where the family needed to cross a moderately busy street walking to our destination. We needed to cross to pre-position ourselves for a left turn up ahead. I was in front and looked back over my shoulder and behind and gauged that there was enough of a gap that two or three of us could cross. The street was busy enough that we would have had to wait a long time to find a gap in the traffic that would allow all of us to cross at once, so we would have to cross in two groups. The first two of the family followed my cue and then Mom just blindly followed them, forcing the traffic to slow and sound their horns. This really scared me. I told Mom that she has to pay attention, as drivers here do not always yield to pedestrians! Mom said she was not paying attention, just following the person in front of her. I guess that logic would have prevented Dad from getting lost but Mom's approach would have made Mom pretty helpless if she ever did get separated from the group. After that, I just waited for a large enough traffic gap or got everyone in a close group and explained the tight crossing we were about to attempt.

The trip from Niamey to Maradi with the family must have been uneventful. I had done that route quite a few times and have no memory of it. Once at Maradi, I introduced my family to some of the staff at the Maradi office, particularly Dr. Ali, the project director. As a project member, I had an official paper giving me permission to travel through most of the northern two-thirds of Niger (an order de mission). However, for my parents and siblings, we would need authorization to visit my workplace at the Dakoro Ranch. We provided names and expected arrival and departure dates to Dr. Ali and waited for the needed signatures through the governor (Préfet) and the livestock service officials. After a couple of days of waiting we were informed by Dr. Ali that permission was granted for my family to travel to Maradi, where I was officially assigned. This was devastating! I only went to Maradi to get supplies and spent 90% of my time at Ranch Dakoro or on a project field trip in the pastoral zone. It seemed that Préfet did not feel that going north into the pastoral zone was safe.

It just happened that Bud Rice and Adamou where also in Maradi getting supplies for their next trip up into the pastoral zone. Bud pointed out that the Préfet had said we could not go to Ranch Dakoro, but did not say we could not go the Ibecetene Ranch where other Peace Corps volunteers were running a grazing trial study identical to what I was doing at the Dakoro Ranch. Bud further volunteered to take us to Ibecetene and back. He had some business at Ibecetene as I recall. Bud had a Scout pickup like we had at Ranch Dakoro, but Bud had removed the gas tank to improve ground clearance since the custom over-sized gas tanks hung low behind the rear axle. He replaced the gas tank with two 50-gallon barrels standing right behind the cab in the back of the pickup. He moved his fuel line into a small opening on the screw-on cap on the barrel's top. This was a bit dangerous if there was a vehicle accident, but it had the advantages of better ground clearance (helped to not get stuck in deep sand or a narrow gully) and only having to fill up with gas once a month. Bud and Adamou essentially lived out of that truck, so there was a lot of gear in the back. Nonetheless, we put Mom and Darlene in the cab with Bud and Adamou; Allan, Dad, and I rode in the pickup box. It was surprisingly comfortable as we arranged the gear (ours, Bud's, and Adamou's) such that there were pseudo seating spots with a view. Adamou and I were front and center right behind the fuel barrels. We used turbans to both minimize exposure to the sun as well as to reduce dust inhalation. I think Allan and Dad might have had a baseball cap and bandana arrangement.

We departed Maradi for Ibecetene quietly, not informing anyone on the project where we were going and headed up a gravel road passing through Bouza and Keita. There were road checkpoints along the way with police checking Bud's paperwork or just waving him through as they had seen lots of the project's white Scout vehicles with our generic permissions to the northern two-thirds of Niger. As we passed through one of the road check stations at either Bouza or Keita, I saw one of my fellow Peace Corps volunteers, Wayne, who came in at the same time as I did. Wayne was talking with the policeman running the road check station. I only had time to wave and say "Hi" to Wayne before we were off again.

We found out later that when Paul Daily, the USAID administrator of the project, heard about our denied permission to go the Ranch Dakoro, he let the Préfet know that he would pull all of the project staff out of the pastoral zone if it was not safe. The Préfet backtracked and approved our permission to travel to Ranch Dakoro, but by then we were traveling "incognito" on our way to the Ibecetene Ranch. This must have put Dr. Ali in a difficult situation not knowing where his staff was, but he never complained about it to me. Apparently, there were radio messages sent out to the police and both of the ranches to roll out the red carpet for the Wylie family. We only learned of this when we returned to Mardi.

Bud was unpredictable and seemed to like leaving the folks traveling with him in suspense of what he was doing. Somewhere between Bouza and Keita, Bud slowed down, pulled off the road, and started driving through a harvested millet field. All that was left of last fall's harvest was the dry millet stocks. It was a little bumpy but there were no formal crop rows, as the planting and cultivation is done by hand.

Slowly, Bud rolled to a stop and got out to announce that we were spending the night here. It looked like the least probable place to camp on earth to me. Mom was a little distraught but did not complain, at least not that I heard. Some went in search of fire wood and others got the cooking stuff and personal gear unloaded. Soon a fire was going and a meal was cooking (usually rice or macaroni with some kind of sauce). Given that we were just coming from Maradi, we probably had a fresh cut of meat along. When supper was over there were conversations about the day's travels, questions about what my family saw that day, and cultural questions. It was pretty dark by the time folks headed off to their mats and blankets for the night. I was tired and slept remarkably well, as I suspect the others did as well.

The next morning was cool. After hot coffee around the fire with a light breakfast, we were off. The Ibecetene Ranch road was a two-track dirt road off the paved road from Tahoua to Agadez. As we entered the Ibecetene Ranch, the road passed through the ranch headquarters. Like the Dakoro Ranch headquarters, it was a collection of cement cinder block buildings for housing and office space with corrals and a large water tank on the top of the nearby hill. I believe we just drove slowly through and waved at some of the livestock staff working there. Routinely, we did not stop at the ranch headquarters, but would see them out at the second well when they came out or when we had some official business with them. Also, this trip was kind of under the radar and we wanted a low profile. Thus, even if the Ibecetene Ranch had gotten the radio message about rolling out the red carpet, they might not have realized that it was me and my family who were visiting at the second well.

At the second well, the Peace Corps volunteers who lived there and the grazing trial supervisor (David Blaine) welcomed us and happily included us in their meals and activities. One of the volunteers at this grazing trial, Roy, was on vacation. He had a white camel called "Lightning," but the Tomacheque (the language of the Toureg) word for lightning. One of the herders saddled up and brought Roy's camel by to give rides. Dad volunteered right off the bat. As Dad climbed aboard, I warned him that the saddle could easily slide forward and that it is paramount that the rider keep his foot on the camel's neck, particularly in the forward and backward lunges as the camel stands up, or a fall is nearly certain for the rider. I suggested that he kind of overlap his heels (shoes or sandals are off at this point) and splay his toes off perpendicular to the camel's neck. This give the rider a wider "foot target" to keep or get back on that neck when the camel stands. Dad said he was good to go. As I stood back, the herder asked if anyone else wanted to ride on back behind the camel's hump. There was just silence, then the herder said, "Bruce?" I said, "Sure!" and jumped on grasping the back of the camel saddle that Dad was sitting on. My grip on the saddle further stabilized the saddle from sliding forward. Finally, Lightning stood up and we were off at a slow walk.

The herders and their families all got to know my family members rather well. These family members still remember some of the herders' names to this day. It was relaxing, fun, and memorable at Ibecetene, both for my family members and for me.

Honestly, I do not remember if Bud took us back to Maradi or if he took us straight to the capital city of Niamey. But we had several days in Niamey before my family flew home. I took them to the zoo in Niamey, not so much to see the animals there, but to see the local crafts and artists' products. The zoo was an exhausting day and some of the artists were very aggressive with their sales tactics. I think my family only bought a few small things.

Another event that I remember vividly was when we stopped at a local meat market to get something for supper. The meat vendor was impressed with my Housa speaking skills but nonetheless was resistant to come down much in price. He had some questions which I was translating back and forth with my family. As I recall, I was tired, hot, and thirsty. In the conversation that I was translating, I started to get weird looks from the meat vendor. I then turned to my family and translated what the butcher had said and my sister just yelled my name. I paused and realized that I had been talking English to the butcher and Housa to my family. Somehow my language toggle switch had gotten confused. I am not sure how long this had been going on, but I suspect just on the last set of translations. I was soon back at the conversation, negotiating for a fly-covered piece of meat (it was an open-air market and all the meat had flies swarming around.). The butcher, I suspect in an attempt to test my local language skills, began making racy comments and jokes which I refused to translate. I told him he had no shame but he just kept smiling and talking. We did finally buy the meat, but I wish I had not as his behavior was simply out of bounds for both of our cultures.

On my family's departure date, we arrived early at the airport, just to be sure they got on the plane, as I had heard that sometimes folks would have trouble getting on their flights leaving Niamey. When they checked in at the ticket desk, they were told that they had not re-confirmed their reservations 24 hours in advance. Dad said they confirmed the reservations when they bought the tickets back in the U.S. The ticket agent pointed out some small text on the printed ticket saying that a 24-hour confirmation was required. They were put on the waiting list but were told the flight was pretty full. Out of the blue, Paul Daily, the USAID project administrator, walked up to say his good-byes and ask how their trip was. He was the one that pushed for our approval to go to Ranch Dakoro, and I suspect he just wanted to hear how everything went on their visit. They updated him briefly about the wonderful time they had, but quickly jumped to the big issue at hand. We waited and waited for them to be called up over the speaker system to get their standby tickets, but none was coming and time was running out. Paul Daily disappeared and within 15 minutes or so my family's names were called and they were granted tickets. They thanked Paul for his help and boarded the plane home.

It was weeks later that I learned from some other USAID or embassy person that Paul Daily had gone to the airport manager and plopped his diplomatic passport on his desk stating that my family would be getting tickets home. I never did get the opportunity to thank Paul after learning this. I ran in the circles out in the bush and Paul in the Niamey circles. I would see Paul only on big project trips or reviews. I also think Paul accepted a position in another country soon after this. Anyway, I and my family are deeply indebted to Paul for his effort both to see that my family could be allowed to visit me and that they could leave at their scheduled time.

I was sad to see my family leave and glad they had a good time. But at the same time, I was glad to see the "drama" frequency drop precipitously as I returned to my normal work issues and problems.

— 9 —

BUSH FIRE (1981)

It was Tabaskay at Dakoro Ranch, and Dana, Jay, and I had bought a sheep to slaughter for the traditional feast and meat sharing associated with that holiday. The previous night I had visited the two French guys who were at the second well. There was an older and ornery guy and a younger, active, optimistic fellow who were building a tick bath so the cattle could be cleared of ticks. When the tick bath was finally complete and filled with water, they invited me to initiate the tick bath with them (no tick treatments applied, just water), so we dove in and swam in the miniature swimming pool. They were talking about going out and getting their "goat" for tomorrow's Ramadan celebration. It's a Muslim tradition to slaughter some animal and then share the meat with neighbors and friends. Most folks were going to do what they could. There was only us three Peace Corps volunteers, our four herders and their families, and the two French tick bath construction crew out at the second well at that time. At the first well was the ranch headquarters. It was a small town with government civil servants (mostly livestock agents) and hired help. They would have electricity at night; the diesel engine used to pump the water would also generate electricity. The pump engine only ran when the large water tank on the hill was low. The second well just had the pump house and watering troughs and they would move the cattle to the second well usually in the rainy season kind of as a grazing management plan. Our grazing trials were at the second well so as to have low impact on the ranch headquarters' operation.

The Peace Corps volunteers lived in huts that were thin circular mud walls up to about waist high and then an arching grass thatch roof placed on top of the circular mud wall. Mine was the smallest, the first built, and was only about seven to eight feet in diameter. Dana's and Jay's were two to three feet larger in diameter than mine. A fourth grass thatched hut served as a kitchen. It had a stove, but had a refrigerator only much later. For the first year, all had dirt floors but the Peace Corps office in the capital city of Niamey insisted that we get cement floors. We argued (dirt floors were easy to maintain) but the Peace Corps doctor forced us to put a thin cement floor in.

Jay, Dana, and I carried four legs of our slaughtered sheep to the four camps where our herders lived. Our herders helped take the cattle in our grazing trials to and from water daily and with various other tasks. We had two Toureg and two Wadabe Fulani herders. The Wadabe are animist with traditional medicines, ghosts, and magic all playing significant roles in their beliefs. We were dumbfounded when Bermu, one of the Wadabe, saw us approaching his house (looked like a small dome tent made out of arching branches covered by black plastic or pieces of woven plastic gunny sack equivalent) with a leg of lamb and told us to stop. He was waving his hands frantically and yelling as if a significant catastrophe was impending. When he got to us he explained that if the ghosts saw us bringing a gift to his house, it could bring bad luck to his household. So, he took the leg and hid it behind a handful of grass and carried it along his thigh as if he was trying to keep if from view. He kindly invited us to his house to chat for a while. After a brief conversation and greeting his wife, we had to depart to cook our remaining portions of the sheep.

Jay always was the cook amongst us three Peace Corps bachelors. We were discussing something (maybe the peculiar antics of Bermu?) when the two Frenchmen came flying up from the well in their Toyota pickup. They were half way out of the pickup before it even stopped, and they informed us there was a bush fire and that we needed to attack the fire. We looked to the southwest and sure enough there was a building column of smoke. We took off in our Scout pickup (Jay was always our self-appointed chauffeur as well) with the younger Frenchman to pick up our herders while the older Frenchman headed back to his tick bath project.

Our herders rapidly hopped in the pickup and we were racing across the tall dry grassland (an annual grass, *Cenchrus biflorus* or Cram Cram, which dies and dries after producing a seed covered in short spines at the end of the rainy season). This was the early part of the dry season so we still had lots of thick standing grass about knee high or higher. The young Frenchman was shouting orders, insisting that we drive to the head of the fire to stop it. I had fought fires with the U.S. Forest Service, putting myself though college. We always attacked a grass fire from downwind, beating it out as you proceed up the fire line. But the young Frenchman was persistent and loud, so we obliged and attacked the head of the fire. As we stopped near the head of the fire we could feel the heat, smell the smoke, and hear the soft roar of the fire in the breeze. Once we were out, Jay headed back to the ranch headquarters to get another load of people to help fight the fire.

We attacked the head of the fire with vigor and good intentions and had put out a 20-meter stretch at the head of the fire. However, the flaks of the fire kept burning downwind and soon had passed our attack position. No one said anything as we walked in the burned area watching the fire proceed WAY beyond our position and burn the unburned portion of our short fire line. The young Frenchman looked particularly dejected with his eye focused on the ground about one meter in front of his feet.

Then one of the herders pointed out Jay walking toward us from the right flank of the fire. I headed over and Jay said the rear leaf spring of the Scout had broken. Jay said the truck would be fine as it was in an area with lower grass and was not in danger of burning. I insisted on seeing it. Jay described that the pickup box and rear fender were resting on the top of the rear tire. Driving it would mean shredding the tire. That Scout was our only means of transportation as we lived literally in the middle of nowhere.

Jay walked me over a little hill and down to an intermittent pond area which was now only dry hard clay. The hard clay soil meant less grass but there still was adequate fuel to carry a grass fire. As I sat on my haunches beside that rear tire, I could hear the soft roar of the fire getting louder as it approached. I glanced up and saw the fire was slowly descending the hill toward us. So, I told myself: 1) you've got 1-2 minutes to come up with a plan and execute it to save the truck, and 2) FOCUS on solving the problem and do not to panic. What could I possibly do? What kind of tools did I have? The situation looked very bleak and insurmountable. Then out of nowhere an idea! I yelled at Jay to grab the jack out of the cab. I stuck the jack on top of the rear axle with the jack's top pushing on the bottom of the pickup box. Then we started turning the crank on the jack. Jay had only one arm but he was turning that jack crank handle fast while I held the jack in position. Soon the pickup box began to lift off the tire. Once we had cleared the tire, our herders had joined us since the fire essentially had burned right around them. We had them knock down the fire right behind the pickup about 10 m long and Jay backed the Scout into the ash of the burned grass. Then I exhaled and celebrated.

We regrouped and began walking back to where the fire apparently started, secured a downwind fire line, and began working up the flanks of the fire. Jay limped off with the Scout driving very slow and careful. He went all the way to headquarters, stopping whenever the jack fell off and reinserting it. After the headquarters truck brought multiple loads of people to fight the fire, Jay brought some of our 5-gallon water jugs in our limping Scout and then headed home. I carried two 5-gallon jugs along the fire line so people would come and drink water and then head off again to do battle. Let me assure you lugging 10 gallons of water around everywhere is not easy. Then it occurred to me the more they drank, the lighter my load. In an hour or two the water bottles were empty. We got resupplied with water and fought the fire well into the night. I did get a picture of Dana fighting the fire in the dark where he was silhouetted against the flame.

Dana fighting fire.

When we got home we were all beat, really beat. We were sore for several days. There was an investigation by the sheriff as to the cause of the fire. Everyone expected some kids trying to smoke a ground squirrel out or something, but the sheriff said they found vehicle tracks near the start of the fire. They indicated a turning and accelerating vehicle along with some gazelle tracks. The sheriff said the vehicle tires were much narrower than the fat tires on our Scout, and the Sheriff suspected the Frenchmen at the tick bath. I mentioned that they had said they were going to get a "goat" that day.

Several evenings later, I went down to the well to bathe and stopped by the tick bath to see what was up. The young Frenchman had left and only the older guy was there, obviously very drunk. He kept pointing at the glow of the ranch headquarters light near the eastern horizon, saying it was a vehicle wandering around and was probably the one that started the fire. It just looked like the lights of the ranch headquarters which we could see glow in the night sky when they turned their generator on. I was disillusioned by his behavior and tried to find a graceful exit and leave.

— 10 —

TO MILK A COW (1981)

Growing up in the Sandhills of Nebraska, I was enamored with calf roping. I rode and took care of several horses and practiced with a lariat in our backyard. While working at a government ranch in Niger, someone (the range consultant from the States?) gave me a lariat. I would practice roping tree branches and bushes while we monitored the milking and measured the cows' milk each morning and evening. Every 6 months, at the end of one of our grazing seasons (rainy and dry), we would get new heifers and milk cows from the ranch headquarters. The ranch director let us pick the heifers for our grazing trial as well as cows with calves for our milk data collection. One of us, I do not remember who, selected this one skinny cow with a crazy look in her eye. I saw several of our herders exchange concerned looks, so it must have been one of us three Peace Corps volunteers, Jay, Dana, or myself. Moukhamadoune, who was one of the best herders we had, seemed to see this crazy cow as a challenge and said that she would be fine.

We herded our new research animals out to the second well where our grazing trials were and where we lived and worked with our herders. We all kept a keen eye on the crazy cow, who tagged along with the herd, but who always seemed to be scanning the horizon and us. After several days at the grazing trials, the crazy cow had yet to be milked and would only let her milk down for her calf. The herders were discussing ways to force her to let down her milk, but had a hard time capturing her and had encouraged me to rope her. On the spur of the moment, when the crazy cow was standing there staring me down, I tossed the lariat over the cow's head and pulled the loop tight around her neck. SUCCESS!! I had worried that if I missed, she would never let me within rope range again. She tried to run away for a couple of seconds, but as the noose quickly tightened and I dug my heels in, she did a quick 180 and charged me. I was running away toward a tree which I thought I could climb pretty fast given the situation I was in. The four herders were all yelling and whooping it up after setting down their milking bowls and running to my rescue. I think Jay and Dana were just laughing, but I really was not focused on them at all.

Moukhamadoune was able to distract her by grabbing her tail and pulling her sideways. I paused halfway up the tree to watch as the crazy cow now turned her attention on Moukhamadoune. He was like a matador, dodging, and with the other herders grabbing the lariat they finally surrounded her and got hobbles on both her front and back legs. She was REALLY MAD now! How could they possibly milk an irate cow? They tried bringing her calf up to suckle, but she would not let down her milk even for her own calf.

This seemed to just "up the ante" with Moukhamadoune. He was going to milk that cow one way or another! There was a discussion between Abdormahamed and Moukhamadoune in Tomacheque which none of us American boys understood, but they quickly sprang into action. Two herders held the cow so she could not move, the third got a milking bowl ready, and (this is the honest truth) Moukhamadoune started blowing air up the rectum of the cow. He would close the opening tight when he came up for another breath of air, by overlapping the skin at the opening. There was cow shit all around his face and mouth, but he was happy, laughing, and determined. Seven or ten blows of air and the cow started arching her back and stomping with her back legs. I, and I suspect neither did Dana or Jay, had no idea what was going to happen. The word was given and the guy with the milking bowl started milking as fast as he could and low and behold, there was milk. After the milking, the cow was released and we all watched her closely, not sure if she would charge or leave. She gave us all a stern look and I was sure she would be headed our way, but then she turned and left.

Moukhamadoune was given a double portion of milk for his family that time, as he tried to wipe the cow shit from his face and mouth. We gave him some water to wash up a bit with.

The crazy cow bolted on the next trip from the grazing trial pastures to the well for water. We thought she would return to her calf at the next milking, but she did not show. After she missed two milkings, we decided we would go after her. One of the herders had seen her north of the grazing trial fence, so two of the herders got in the back of the pickup with me and my lariat. Jay was driving and Dana was riding shotgun. The herders' plan was for me to rope the cow from the back of the moving pickup while bouncing across the prairie. We found the cow and chased her with the pickup over rough ground. The ride in the back of the Scout required we hang on with both hands. As we got closer, she seemed to realize she could not outrun us and was slowing down. I grabbed the lariat and hoped I did not fall out of the back of the truck. Moukhamadoune and Abdormahamed were each hanging on with one hand and with the other trying to hold my feet down on the floor of the pickup box. I made my toss and . . . again, PAY DIRT! I was relieved as I had worried if I missed, the rope could get tangled on the rear tire and axle.

So now I had a wild crazy cow on the end of my rope . . . AGAIN! As Jay slowed the truck to a stop, the two herders in the back with me helped me pull the rope and keep the cow from running off. Again, she stopped, looked us over, and then charged the pickup! WHAM! She actually put a little dent in the side of our pickup just below where I was standing in the pickup box. But we quickly reeled in all the slack rope and she found herself short tethered to the side of our truck.

Jay drove SLOWLY back to the grazing trials and we let her calf drink milk. She seemed to pay the calf little attention and was looking off to the distant bush. We put her back in the grazing trial fence, but at the next watering, she ran off again. We just let her go and told the ranch herders and director that she was loose. She rejoined one of the headquarter herds back near the ranch headquarters several days later, but always remained spooky.

One night many years later, while working in Alaska, the conversation around the fire turned to the question of if one could possibly milk a moose. I said, "Yes, definitely," and recounted the crazy cow story. One of the folk around the fire was a young female graduate student. She did not like my story at all! All I could say was that it was true.

— 11 —

BROKEN TRUCK (1980)

Dana and I were doing errands and reading around our huts. Jay had taken the truck to run an errand, insisting he go alone. Suddenly, Jay stepped out from behind my house, startling both me and Dana. We asked where the truck was and Jay said it was three-fourths of the way to Fuko. Fuko was an abandoned Hausa village that was just within the southern border fence of the Dakoro government ranch. The ranch had recently forced the Fuko population to vacate their village a couple of years before. Fuko was probably about 10 km or so away from our huts on a lightly used road that cut diagonally from the second well (where our grazing trials were) to the main road going south toward the small town of In-Tuilla (12 km south of the ranch's southern fence) and the bigger town of Dakoro (another 20 km further south).

Jay said the truck had just broke down on him and he was certain it was the carburetor. Astoundingly, he also asserted that I could fix it! I asked, "How?" I was dubious that he even knew what was wrong with the truck. He said he and I would walk back to the truck with my tools (not much, just a couple of screwdrivers, a pliers, and a crescent wrench) the next morning. I asked if the battery was dead and he assured me that the battery was still charged (he had not continued to try and start it until the battery was dead). What the heck, I would give it a try, but as I understood, carburetors were complex and tricky. For that reason, I had never tinkered with them much on my own vehicles or lawn mowers I had owned or worked on for Dad.

Jay and I got to the truck by mid-morning, as the temperature was still climbing. I climbed in under the open hood, where I could access the carburetor easily. I was going to disassemble the carburetor slowly and carefully, being sure not to lose any parts. Off came the front plate to the carburetor, revealing the insides of the small float tank with a jumbled mess of parts consisting of a small axle, a float, a spring, and other odd assorted pieces. Jay was right in his diagnosis that the carburetor was the issue after all!

I carefully removed all the loose pieces and was completely baffled by the assortment of pieces. I was pretty sure that I would not be able to figure out this reassembly puzzle. I started looking closely at the pieces and trying to see how they possibly fit together in a functional way. I noticed a spot on the inside of the float front cover which I had removed. There was a rubbed spot on the inside. I assumed that it had to be the float that rubbed there. Sure enough, there was also a rubbed spot on the float. This key helped immensely as I now knew how the float should be positioned. Suddenly, I knew how all the pieces fit together! I rapidly started re-assembly with Jay handing me the various pieces and tools I needed. Once it was assembled, Jay tried to start the truck. It only turned over without firing. He tried it again and it fired right up. I was stunned and Jay was in a celebratory, cocky mood. I grabbed my tools and jumped in the passenger door and in no time, we were back at our huts.

We had lots of vehicle issues and I seemed to be assigned "mechanic" at the second well after that.

— 12 —

TRIP HOME FROM THE PEACE CORPS (1981)

I was in the capital city of Niamey, finalizing my completion of my Peace Corps tour of duty (roughly 2.5 years for non-school teaching assignments). The terminology that was used for this was "termination." I do not remember all the forms and interviews I had to do as part of my termination process, but I do remember being focused on the planning of my trip home. I had two options: 1) Peace Corps provides an airline ticket home, or 2) get the cash equivalent and find my own way home and explore the world. I chose the latter as I had heard stories of an earlier Peace Corps volunteer who had crossed the Sahara Desert overland on his return home. He was robbed of everything by some asshole in the southernmost city in Algeria, Tamanrasset. He was able to beg his way to France and then got a job picking grapes in southern France. He soon saved more money than he had saved in 2 years in the Peace Corps and had greatly improved his French.

I sought a visa to Algeria and after the interview I was told I would need to show proof of a purchased exit transportation out of Algeria. They apparently did not want me to be a burden on the local population. It was kind of a kick in the gut that they did not think I had either the funds or wanted to avoid having to beg after I got robbed. So, I bought an airline ticket from Algiers, the capital city of Algeria on the Mediterranean. With my Algerian visa, I was preparing my departure and had a date fixed. Most of the Niamey Peace Corps volunteers were aware of my foolish plan and one introduced me to a fellow who also was traveling to Europe by land. This fellow was one of the sons of the ex-president of Liberia. His father had been killed in the recent coupe where Sargent Doe had recently taken over the Liberian government. I do not remember his name so let's call him Randy. Randy only spoke English, so I was to serve as his translator for Housa and French. I, too, was a bit comforted with having a traveling companion as well. Since Randy was much bigger and stronger than me, I hoped he would act as my bouncer, if needed. We agreed on the departure date and the date when I needed to be in Algiers to catch my flight. We set a rendezvous time and place.

My next task was that I wanted to send money ahead to a French bank, where I could pick up the necessary funds to travel across France, visiting my cousin Lisa who was in Strasburg, France. I would then fly out from Brussels, Belgium, to my parents' home, which was now at Wise River, Montana. I went to a bank in Niamey and then we selected a bank in Marseille on France's southern Mediterranean coast. This was, I thought, a clever way to minimize the impact of potentially getting robbed in Algeria.

As I recall, Randy and I boarded the SNTN, which was a fairly modern state-run one-day bus service to Agadez in northern Niger. It was a little bit spendier than a series of bush taxis to leapfrog our way to Agadez in several days. Bush taxis are often poorly maintained and severely overloaded and crowded as well. It was a long day's travel, but I was familiar with the route as I had done it on several of our project trips and on my Peace Corps vacation. The route extended northeast through Tahoua, which was a rough northern limit of the farming zone and the southern limit of the pastoral zone where livestock herding is the dominant land use. As we proceeded northward, the climate became drier and drier, transitioning from Sahel scattered trees in a grassland to more desert vegetation with rare trees and shrubs and only patches of grass.

Agadez was one of my favorite cities in Niger. I liked its mud adobe architecture where they were creative in many of their buildings and its rich history. Historically it was a major city for cross-Saharan commerce and salt caravans, dominated by Toureg history. It is probably most famous for its mosque, which has a tall adobe tower for announcing prayer time. Randy and I must have split a hotel room. The next day Randy and I went down to the bush taxi station and asked about transportation for two for Tamanrasset, Algeria. We were asked to wait a bit, so we killed some time chatting with folks. Soon word came back from the bush taxi rumor mill that there was a truck headed to Tamanrasset in two days. We met with the driver, El Hadji, and confirmed the price, the number of days for the crossing (3), supplies we would need (sleeping gear, four liter or greater water bottle; the driver would refill it as needed), and departure time and place. As I recall the driver provided basic meals for breakfast and supper.

Our departure was a cold morning in the cold season (January or so). Our truck was a large Mercedes dump truck with duallies in the back. There was a dozen or so of us in the back of the empty truck bed. There must have been some gear piled toward the front that we all sat on looking forward. I had anticipated seeing the town of Arlit where there was a uranium mine, but I was informed that we would be taking another route. The Arlit route was for tourists and had a lot of loose sand, whereas the truck route followed lowlands with heavier soils west of Arlit. I enjoyed talking with El Hadji, the driver. He was curious about what I had done in Niger, and I was curious about what he was doing. His route was to go up to Tamanrasset empty, load up with dates, and then return to Agadez and Abalak by Ruwa (small town northeast of Tahoua). My bush path route between the two government ranches with the cattle scale in tow would pass through Abalak when I returned to the project. Since I had verbally been assured of being rehired by the Livestock project I had worked for as a Peace Volunteer, I suggested that someday in the future, we might meet in Abalak. We both knew the odds of that happening were very low but enjoyed the possibility.

During day two, we had to go through a sandy stretch so we all dismounted and watched as El Hadji's crew of two deployed the sand ladders in front of the rear tires. Progress was slow but sure and the distance was only 100 m or so. We also came to the Algeria and Niger border crossing.

There were just two small adobe buildings out on a sandy plain, one for the Algerians and one for the Nigeriens. There were, however, lots of vehicles with about 20 trucks and 30 cars and motorcycles waiting to be processed. I had no problem with my passport and assisted Randy with translation with the customs officer. I hung close to our truck as I did not want to be left behind. Randy, on the other hand, wandered off to visit with the tourists. I suspect we was telling his story in hopes of getting some donations. I asked him how he was going to finance his trip and he said he was a traditional dancer. He had an elephant rib in his gear that was part of his routine. He seemed very confident of his ability to earn money this way any time and any place.

Suddenly, I heard someone yell "Fire!" and turned to see flames near the battery of El Hadji's truck, which was also close to the diesel tanks. El Hadji yelled for folks to throw sand on it. There were about seven or so of us throwing sand. The problem was that right there, there was a small amount of gravel mixed in with the sand. I thought it more important to get the flames out than sort out the rocks and slung my two hands worth as hard as I could at the base of the flames. One could hear the rocks from my "sand" bouncing off the truck. I saw El Hadji stare at me out of the corner of my eye as I re-loaded with "sand," but he said nothing. Soon the fire was out and El Hadji and his crew rushed in to do repairs. Once again, we loaded up and I noticed Randy was not here. I was not about to go off looking for him and miss my ride, but El Hadji insisted I go get him, pointing out where he was and assuring me he would wait for me. I ran up to Randy, who was talking to a motorcyclist, and yelled, "We are leaving! If you want to go, get in the truck!" I then did a 180 and took off running back to the truck. Randy caught up with me, got in front, and turned around running backwards facing me and asked me not to be mad at him. I said, "I am not your mother," and sprinted around him. I was embarrassed about causing the delay and may have mumbled some unkind things in Housa.

The second night out of Agadez, we heard ostriches calling in the early morning.

The third evening Randy and I were dropped off at the bus station in Tamanrasset. We said goodbye, which was especially tough for me with El Hadji. When I did return to Niger, I always looked for El Hadji's truck when I passed through Abalak. I made an effort to always stop and get something to eat there. Once I was leaning against my white Scout pickup, eating a brochette when some guy walked up to me. I looked up and it was El Hadji! I hugged him. This is way beyond the bounds of normal conduct in Niger, but El Hadji did not resist. I asked him where his truck was and he said he was driving a different truck. I had pictures of our trip to Tamanrasset in his truck that I rode in and gave him the ones he was in. El Hadji had told the folks he was with that he knew that white guy over there and they did not believe him. I think it was profoundly evident to everyone that I knew El Hadji very well.

Randy and I got a bite to eat in Tamanrasset and the thought occurred to me that now I had to switch from Housa to French. My Housa was stronger than my French, but my French was quite passable. Randy and I slept on the ground in the bus station, since the bus going north left at 5 in the morning. We were tired and I rolled my pants up, using them as my pillow. I recalled that this is where the other Peace Corps volunteer was robbed and vowed to be vigilant. It was hard ground, but thanks to fatigue, we both were soon sound asleep. I awoke to a car idling with its headlights on and some guy trying to remove my pants from under my head. My mind was a little groggy, but I soon jumped into action, confronting the guy and asking him what he was doing. Randy was soon awake as well. The thief just said that he was sorry and had mistaken me for one of his friends. I did not believe it but was not alert enough to say, "Yeah, right!" I was just glad there was no physical confrontation. It was comforting that Randy was a big, strong guy though.

The bus showed up early in the morning and we climbed aboard to be welcomed by clean and comfortable upholstered seats. We learned our destination for the day was In Salah, where we would overnight again. This time Randy and I found a tourist campground. It had a cement cinder block wall surrounding it with a large gate. On the inside of the perimeter wall were cinder block huts with large doorways but no doors and cement floors. Randy and I paid for one of these. As the sun set and darkness settled in, I was not quite ready for sleep and wanted to look out over the camp area from our doorway. I was kind of curious about the mixture of folk that were here. There were mainly Europeans headed south to visit Africa. Randy and I were discussing something, probably our destination for tomorrow on the bus, when a dome light in a small station wagon nearby turned on and we could see an attractive European young woman sitting there, but constantly moving and readjusting her sitting position. I thought maybe some kind of exercise or something, but Randy calmly pointed out that she was having sex. How naïve I felt! It seemed I was always informing or helping Randy, but here he corrected me. Soon she stopped and shut off the dome light. In the same nanosecond Randy and I simultaneously took one step back into the darkness of our hut and hit the hay.

In the cold morning Randy headed off to a table full of tourists drinking coffee and eating bread and jam. He returned to say he had chatted with the gal in the station wagon. Per Randy she was quite nice and attractive.

We were soon off to the bus station and boarding the bus north to Ghardaria. Some of the folk on the northbound bus started to look familiar. A middle-aged, balding, Arab man sitting behind us started asking us where we were headed and where we were from. My responses about myself were short and implied a boring science-based profession in Niger, a sure conversation killer. Randy, however, launched off on the story of his father being killed in the coupe in Liberia and how he had escaped with his life. Randy alluded to a life on the run, looking for some assistance and help. Bear in mind that I was doing all the translating so I got sucked into the conversation that lasted hours. When we arrived in our destination town of Ghardaria, the middle-aged Arab invited us to spend the night at his home. I was a bit suspicious but Randy was happy to agree.

We walked with the man through a maze of streets in a middle income residential area in the darkening dusk. Just as full night descended on us, the Arab man stopped and pounded on a big blue metal gate. A woman's voice answered and unlatched the large car-wide double doors. The Arab man proceeded to open the doors and we caught a glimpse of a woman's shawl ends just as the ends of a flowing gown disappearing around a dark corner. We followed the Arab man upstairs to a cement room where we sat on a mat on the floor. His wife came to the door and talked to the Arab man while hiding behind the door. Then she slid a tray full of tea makings (tea, sugar, water, and an electric hot plate) onto the floor from behind the door and was gone. The Arab man prepared tea as Randy continued his descriptions of the sincerity and seriousness of his situation, with my translations. Soon a meal of couscous and meat on a tray was slid onto the floor from behind the door. The Arab man's wife said something to him, who then said his wife welcomed us and hoped we would enjoy our stay. As time grew late and my translation skills faded, the Arab man excused himself and returned with a rolled wad of Algerian Dinars as big in circumference as a grapefruit and gave it to Randy to help him on his way tomorrow. I was completely stunned and said so, stating that the Arab man did not have to do that. His response was that because I was an American, I would never understand. I shut up but still had a bad feeling in my stomach as if I had just been part of a scam.

The next morning, Randy and I said our good-byes to the Arab man and hurried off to catch the bus to the capital city of Algiers on the Mediterranean coast. We saw Berber herders' cloth tents in the desert and on the long climb up and over the Atlas Mountains. Algiers was a bustling, busy city. At the bus station, Randy and I quickly found a taxi driver who said he knew about a cheap hotel. We grew suspicious of our taxi driver as we left the city limits. We were dropped at a high-rise tourist resort well out of town. We stayed one night and moved out the next morning. This time we found a taxi driver that did take us to a cheap hotel downtown. We shared a standing toilet with everyone on our floor. The toilet always seemed fifty to me, but the price was right and I felt more at home in the cheaper hotel.

Randy and I headed to the U.S. Consulate in Algiers, where they questioned us together and then separated us. Based on their questions to me when we were separated, the Consulate staff seemed to suspect that

Randy might be somehow forcing me to help him. I assured them that we were traveling companions who crossed the Sahara together, nothing more. I merely wanted to let him state his refugee case to their staff. The U.S. Consulate staff suggested that I catch my flight and head back to the U.S. on my own. They did not grant or promise Randy any refugee status. As I left for the airport to fly to Marseilles, I felt a little reluctance to leave Randy to his own devices (a little like a mother, I thought to myself, which I vehemently did not want to be at the border crossing), but I focused on completing my trip home alive and well.

Marseilles was a bustling modern city full of white folk. Things were pretty clean and on time. But my first priority was to get some American junk food—McDonald's burger and fries perhaps but my highest desire was for pizza. I wandered the streets randomly looking for such places and after several hours of finding none, ducked into a Chinese restaurant. The food was good and hit the spot. Next, I got a hotel room. The next day I checked into a youth hostel and they were sure to point out that their curfew was very strict. The exterior door was locked promptly at 10 p.m. my priority was to get the money I had wired myself from Niger. I gave a taxi driver the address and the name of the bank, Barclays Bank as I recall. The taxi driver pointed to a building and said, "That is the address, but it is not a bank." Puzzled and confused, I stumbled out of the taxi and paid the driver. He pointed out a bank with a different name half a block away and suggested I try there. It sounded like a harebrained idea to me, but I could not come up with a better plan, so I walked into the bank and described my situation. A bank official was called. In his office, they asked for my passport and then stated that they did indeed have my money. I was stunned, and very elated! That night I celebrated at a local bar and underestimated the travel time to the youth hostel and found myself locked out. I headed back toward the town center and stopped to ask a man directions to a reasonable hotel. He was facing away from me, and when he swung around to address me, he was all smiles, an unfamiliar reaction compared to my recent interactions with French locals. But I quickly realized he was so drunk, he could barely stand, so I smiled back and proceeded on. Ultimately, I did find a hotel.

I traveled by train to Strasburg, France, where my cousin, Lisa, was studying. I met her, and since her French was much better than mine (I think I spoke with an African accent and native French speakers were always the hardest to understand), I allowed and encouraged Lisa to do most of the talking. We went to the market and visited some of the tourist destinations in town. Most importantly, we bought my airline tickets to Missoula, Montana, near where my parents now lived in Wise River, Montana. I caught the train to Brussels, Belgium. I had half a day to wander about town, slept a night, and boarded my flight home to the U.S. Once settled on the plane with an English-speaking crew, I felt back in my norm again and very comfortable and relaxed.

— 13 —

AN OFFICE FOR RANCH DAKORO GRAZING TRIALS (1982)

I had returned to Niger as a USAID consultant and had delivered two new Peace Corps volunteers, Lisa and Barbara, to replace Dana, Jay, and me. Adamou, who had help me immensely in my initial period as a volunteer, was already at the Ranch Dakoro grazing trial helping out. Lisa, Barbara, and Adamou all endorsed an idea Jay, Dana, and I had only given lip service to. There was need for a shaded lounging and office area near the kitchen hut. We had been eyeing some of the abandoned huts in the vacated town of Fuko, just within the ranch perimeter fence. I had gotten approval from the ranch director to remove one of the thatched roofs from one of the now vacant huts to use as an office for Lisa, Barbara, and Adamou. The problem was that the size and weight of the thatched roof was more than we could handle, and naturally we wanted the largest and "best condition" thatched roof we could find as long as the price was right. For some reason, Adamou was confident, enthusiastic, and excited that we could: 1) get that thatched roof in the back of our pickup, 2) transport it in one piece, and 3) get it up on short poles so we could duck under easily to get in and out of it while allowing breezes to cool us.

So, we headed off to Fuko with Abdormahamed and Moukhamadoune, two of our herders, in my pickup with a bunch of rope and high ambitions. We backed the truck up to the selected hut and tried to lift the thatched roof off the circular mud walls. But no go! We could not budge it! It was really stuck on there. Maybe rain had allowed the mud walls to essentially become glued to the thatched roof. I rigged a rope under the back edge of the thatched roof and up, over, and around the roof peak, over the pickup cab, and tied it off on the front bumper of the pickup, the idea being when the pickup pulled ahead, it would roll the thatched roof into the pickup box. What did we have to lose? There were at least a dozen thatched roofs to pick from. My primary concern was that somehow the pickup might be damaged or one of us get crushed by the heavy roof.

As the truck inched forward, there were some crackling sounds, and I started to think all we would do was tear the thatched roof in half. But at last it started to tip toward the pickup. I think we stopped and backed the pickup closer to the hut and tightened the rope and gave another tug. The rest of us were trying to guide the thatched roof into the truck, but we were very aware of its weight if we got stuck under it. Luckily, the thatched roof, with lots more crackling, rolled into the truck bed, the pointed peak roughly pointing forward and slightly upward. We tried to further secure the ropes for the journey to the second well. There was some posing for pictures as well. Step one was done!

Adamou (left on top of cab), Moukhamadoune (left on the hood), Abdormahamed (climbing on to the hood), and Lisa (in the cab).

I drove the pickup idled the whole way, stopping once or twice to inspect our load. I think several of our team were walking along both sides of the truck keeping an eye on our precious cargo. Our confidence grew as the roof appeared fairly stable in the pickup box. I think ultimately some our crew piled into the cab with me, with one guy in back standing inside the circumference of the roof, keeping an eye on it. We arrived at our huts at the second well with little fanfare and a lot of SLOW driving. Adamou had instructed us previously as we had prepared a bunch of Y-shaped wooden posts about head tall. We used all of our four herders, Adamou, Lisa, Barbara, and myself to unload the roof and set it right side up on the ground. We used a rope to get the diameter of the thatched roof base by slipping the rope under the edge of the thatched roof and lifting the roof edge as needed. Next, we marked the circumference in the dirt using the rope radius and a center pivot point in the dirt a short distance from the kitchen hut. Post holes were dug and the Y posts inserted into the holes such that the Y was at about chest high and all the posts around the circumference leaning at various angles as they were loose in the holes. We enlisted our four herders again to help with the final heave-ho to lift the thatched roof off the pickup and position it up on top of the Y posts. Then the post holes were filled in with dirt, straightening the posts and packing dirt around them. This allowed us to adjust the angles of the Y posts to best support a semi-level thatched roof. Adamou obviously had experience with this as he gave guidance on many of the installation steps. Boy, were we proud of our shaded lounge! What a nice place to do paperwork (pasture herbage estimates, cattle weights, milk production statistics, which was a fair amount of our work) and casual conversation around meal time. It was both the dining room and our office!

— 14 —

FINAL REPORT (1983)

The USAID project, the Niger Range and Livestock project, was having trouble getting the range consultant to produce a final report. I did not know what the issue was, but I was called down to Niamey to meet the USAID person in charge of the project. He said that without a final report, there would be no range management component in the second phase of the project. The project had a sociologist section with staff working with the Toureg and the Fulani, an economic section, and the range management section. Without the final report the research at the ranches would be closed and all the range management Peace Corps volunteers sent home or re-assigned and the expatriate staff (me) laid off. He wanted me to write the final report for the range management section. I was skeptical that I could do that, but he insisted that I could, with the help of the Peace Corps volunteers at the ranches. I agreed to try.

It was quite emotional telling the volunteers that we had to write the report or go home. Each team of Peace Corps volunteers at each ranch was responsible for writing primarily the results of the grazing trials. This involved analysis of all the monthly heifer weights and our estimates of how much grass was remaining in the pastures at each monthly weighing. Gain per head and gain per hectare were plotted against the remaining grass in each of the three pastures for heavy, moderate, and heavy grazing pressures. This is known as the residual herbage approach to determine when to remove animals from a pasture.

There were other results to summarize from the grazing exclosures around the pastoral zone, where species composition and pasture production with and without grazing were contrasted. We also had other studies to include: clipping studies, the bite counts done on the heifers and a camel or two, and some results from a hay production study. Each ranch prepared their results and then convened at the Ibecetene Ranch to combine results. There was a problem with the Dakoro Ranch gain per hectare and gain per head curves. I re-analyzed the data about 10 times or more with no improvement. Finally, I started including the milk cow weights in with the heifers, as the milk cows were retained in the grazing trial pastures, or maybe I included the calves' weights (who drank the milk), I honestly do not remember. Whatever permutation of cattle weights that I used had logic behind it and resulted in a reasonable set of curves.

I needed one of the volunteers to go down to Niamey and start writing the combined final report. One volunteered. The other volunteers and I continued to wrap up things at the government ranches. In about three weeks, several of the volunteers and I went down to Niamey to help put the final touches and edits on the final report, only to find that minimal progress had been made. So, I was stuck with a report only one-fourth written and a looming deadline. I started writing and turning in sections for the project editor to review. She liked the sections that I produced. I worked hard and finally got a rough draft completed. Several of the volunteers who had come down to Niamey with me also helped review and edit the report, while the volunteer who was supposed to have done the bulk of the writing vanished. The range management section of the report was finalized in time for project review and phase II design team.

I ran into the range consultant who would not write the report back in the U.S. at a Range Society meeting, and he told me he tried adding in the milk cows, but it did not make the curves fit. He basically strongly implied that I fudged the data. He also pointed out how my agreeing to write the final report somehow undercut his credibility or negotiation strategy. Then he walked off just shaking his head in disgust. This hurt deeply, as I had learned a lot from the range consultant and valued his opinion. But, I came to the conclusion that I made good decisions based on what I knew and that I could never please all the people all the time.

— 15 —

STUCK (1982)

I was driving the two-track road with my Scout pickup, pulling the cattle scale trailer behind me. Every month I had to move the scale between the Ibecetene and Dakoro ranch grazing trials to weigh the heifers and milk cows in the grazing trials. It was mid-rainy season and we had just had a decent rain at the Dakoro Ranch. I prepared the scale for transport and hitched it to my pickup. I was kind of a nomad myself, living out of my truck essentially. I had a woven reed mat and a blanket rolled together and tied on top of my pickup cab. I would sleep on the ground by my truck while at Ibecetene or Dakoro Ranches. This trip seemed fairly routine, as the truck and scale appeared in good condition.

The route went north of the Dakoro Ranch about 12 km to the government well at Abouhiya (diesel powered pump). Then, I would turn west on another two-track road to Abalak, which was about an additional 35 km or so. Abalak was a small village on the paved highway (funded by the uranium mine in the northern town of Arlit) from Niamey in the southwest to Agadez in the north-central part of Niger. I think it was about 20 km or so southwest of Abalak where the two-track road turned off to the west for about 15 km to the Ibecetene Ranch grazing trials. I had done this route many times and was perfectly comfortable doing the route alone and interacting with the pastoralists along the way. But this time it was very different.

The significant recent rain made this trip miserable. There were many large puddles on the road, particularly between Abouhiya and Abalak. My predecessor, David, had given me advice, saying he always entered the puddles slowly as this increased the chance that one could back out if it was too deep or too muddy. Also, if at all possible, it was advisable to detour around off road to avoid the puddle. It seemed that I detoured around at least a hundred or more puddles on the road. The trip was taking nearly three times as long as it normally did, and I was getting tired of it. Finally, I got to Abalak and headed southwest on the pavement. As I turned off on the two-track road to Ibecetene, I again encountered numerous puddles that I diligently circumnavigated. I finally arrived at the Peace Corps Volunteers' huts, but found no one. Then I remembered they had relocated from the western edge of the grazing trial fences to the eastern edge, which was closer to the pump station and well. I was tired and it was dark. I was getting fed up with all these puddles causing sometimes major detours off road. I proceeded north along with the western edge of the grazing trial fence, only to encounter a huge muddy area that obviously recently had held water. I was less than a km from the Peace Corps huts, so I decided to roll the dice and take a chance. I put the truck in four low and got a run at it. I got about to the middle of the 60-yard muddy crossing before all four wheels just sank and I was hopelessly high centered in the middle of the mud patch. Nothing to panic about. I grabbed my bed roll and walked the last km or so to the Peace Corps huts. The volunteers and their four herders wanted to go help pull my truck out. I was very tired and more than a little angry, so I suggested we all have a good night's sleep and attack the stuck truck in the morning.

There was one catch though: the U.S. ambassador's son was visiting and helping out at the Ibecetene grazing trials. I was pretty sure he would be telling his Dad all about this. In the morning, we headed off with the Peace Corps volunteers and the four herders in the volunteers' pickup. First, we tried pulling it out with the other truck using 3-4 lengths of barbwire kind of woven together. The volunteer's truck only spun its wheels. So, we began digging, then pulled some more. Dig some more and then pull, dig then pull. This went on for 2 days without any significant success. The headquarters crew happened by with the pickup box on their range rover pickup packed full of people. The ranch driver was confident he could pull my truck out, but I insisted that he pull in unison with the volunteers' pickup. Even that did not work. So, the headquarters crew headed back home, promising to send their tractor out the next day to pull my truck out. On the third day, we continued to dig and pull and dig and pull. My truck now was at a fairly steep angle, with the front being considerably deeper in the hole than the rear. Finally, the tractor from the ranch headquarters showed up and quickly pulled my truck out. We were all happy.

That hole where my truck had been stuck became a watering hole for cattle after a rain and persisted for many years. Bruce's pond. I never did hear any repercussions or feedback from the U.S. ambassador either. Herders were quick to advise me to go around mud holes and not try to sprint across anymore.

— 16 —

FINDING A WIFE (1982)

After coming home to the U.S. after Peace Corps, I was awaiting a phone call or a letter from USAID to discuss the job they had promised me back in Niger. None came so I proceeded to enroll in the University of Montana in a Master's program in Range Management. I did some skiing and was staying temporarily at the house of the former Forest Service Ranger at 9 Mile with his wife and one daughter. I sent a letter to the USAID head of our project in Niger, the Niger Range and Livestock project, letting him know I was skiing in Montana and enrolled in graduate school as I had not heard from him about a job opportunity. I provided my phone contact. Within a week (the mail went via the diplomatic pouch) I got a phone call from the project manager in Niger. He said he was expecting me to return to Niger for the job. I said no, that I would only go to Niger if I had an accepted job offer. This would mean that USAID would pay my travel and shipping of personal effects. Within a week I had a contract in hand for the Range Supervisor on the project, which I promptly signed. I was able to withdraw from grad school and was soon on board a flight back to Niger.

As the Range Supervisor for the Niger Range and Livestock project, I was living out of my truck, driving from the city of Tahoua, to the Ibecetene Ranch, to the Dakoro Ranch, to the southern city of Maradi, and back, once a month. Times in Tahoua and Maradi were spent looking for work supplies, groceries, some updating of the project chief, project meetings, and sometimes in the local bars with Peace Corps and Nigerien friends. Dana, a former Peace Corps volunteer who I had worked with at the Dakoro Ranch, shared a 2-bedroom modern house in Tahoua with me. Both Dana and I worked for the project and spent most of our time out in the bush. Dana and I occasionally overlapped each other in Tahoua.

I had a personal rule that I tried to follow, but ultimately failed. I wanted to avoid having a relationship with any of the women with whom I worked, particularly in the bush. I just did not want the drama and heartache of a fight or break up distracting from work relationships or concerns of favoritism from other work colleagues. I had two Peace Corps girlfriends discreetly, one in Maradi and one in Tahoua. These were just budding relationships.

At Ibecetene Ranch the two Peace Corps volunteers paid the herder families to cook supper for them, and they ate with the herder families. The volunteers rotated their supper destination between their four herders. I think it was every month they would rotate to another herder's family. It was my second or third month on the job as Range Supervisor when the volunteers and I were eating at Seidi's hut. Seidi introduced us to a married woman, Mariama, who was his niece and was visiting with him for a while. The woman had a nearly one-year-old daughter. Seidi had learned how to do our pasture production estimates and random pasture clippings. We used this approach to quantify standing herbage levels in the pastures at the time we weighed the heifers and milk cows in the grazing trials. Seidi also had learned how to play a popular card game there with the Nigerien civil servants called Eight Americans, which was just a game I knew as Crazy Eights (very similar to Uno). The deviation on this game was that it was a speed game, and if you did not quickly proceed when it was your turn, you were sanctioned and had to draw a card. Mariama quickly picked up the game as well, impressing us all with her rapid learning. We would play cards after supper and then head to bed. Mariama struck me as an attractive woman with a very nice personality. There was a lot of joking and kidding around the card games.

One day I was preparing to head down to Tahoua for supplies and reporting to the project chief. There was a small group of folks sending me off consisting of a volunteer, several of the herders, and some of their wives and kids. Mariama called me aside, I assumed for a special request from town. But she floored me when she said she liked me. I responded that she was married, but she said that she was seriously considering proceeding with a divorce as she had caught her husband cheating on her a month ago and left him to come stay with her uncle Seidi. She then proceeded with some request from town. It was, I think, a bolt of black material the women use as wrap-around skirts.

In Tahoua, I suspected Mariama might just be playing me along to get the clothing, but it was not expensive and I decided to play along. I had never had such an approach from any traditional pastoralist woman in Niger. I had heard stories about such things, primarily related to Fulani women, but very rarely Toureg women. As time proceeded, Mariama and I interacted at cards and other group meetings, never alone. After about a month, as dusk harkened, I rolled my reed sleeping mat out near my truck, which was a bit distant from the other huts of volunteers and the herders. I laid down, with my blanket alongside for when it got colder at night along with my water canteen covered in burlap. I was soon dozing off and I was on the cusp of falling asleep when I heard someone whispering and saw two black shapes walking toward me. It was Mariama and her aunt, who was slightly older than Mariama. Mariama claimed to be 19 years old. Mariama, all in whispers as she wanted no one to know she was there, told me about Toureg courtship protocols and invited me to come by Seidi's hut to court her. I was quite taken aback and contemplated this information and offer diligently.

The Toureg courtship protocols seem convoluted to a western perspective, but may be appropriate when all the family live together in tight quarters with little privacy. Single young women and men were frowned upon if they are seen or found alone. It seemed to me that their society went to great measures to ensure that such alone time was minimal and not encouraged. The protocol goes that the male suitor sneaks into the tent or hut of the family of the woman he wanted to court in the dead of night when all were asleep. The woman and the man then would have whispered conversation near her bed with her mother and dad or, in this case, her uncle and aunt a hand's reach away. This goes against normalcy in my western upbringing and I wondered what Seidi or his family would say if he caught me sneaking around inside his hut in the middle of the night. Would he mistake me for a thief and pull his sword or a knife?

The arguments in my head that night continued late into the night. I finally felt that Mariama would not have invited me if this was an imminent threat to me. On the other hand, she was still married and I was sure her husband would not be amused. Her husband lived in Abalak, but he could arrive by camel in the dead of night as well. I very tentatively tiptoed into Seidi's tent very, very late in the night after the moon had set and all was nearly pitch dark. Mariama and my initial conversations were pretty benign, just to get to know each other a bit. This continued for nearly a month, with Mariama always inviting my return. In the day, we hardly made eye contact and only had short conversations. But one night as I was preparing to leave, she grabbed my beard and pulled me close for a kiss. Thus, things began to escalate between us. I would get jealous of her being her friendly self with Nigerien civil servants from the project on visits or on work detail with us at the Ibecetene grazing trials. I only learned 30 or so years later, some of these visits were indeed proposals of marriage to Mariama. Mariama initiated the divorce process with her husband. As I understand it she had to wear a particular necklace for a couple of months. As long as that necklace was not removed from her neck during the time required, her divorce would be granted. The trouble was that her husband would visit occasionally and try and talk her out of it and tried to forcibly break the necklace off. In contrast, it was quite simple for a man to divorce a woman.

From my perspective, I was being run ragged. Work all day and then stay up all night talking. I would catch up on my sleep a little bit when at the Dakoro Ranch, but when there we were busy with cattle weighing and pasture production assessments. The decision before me, as I saw it, was to simply pick the best of my three girlfriends. My approach was to be heavily weighted in logic and facts, weighing advantages and disadvantages. I wanted to give due diligence to Mariama and sought advice and recommendations from two Americans I knew who had a Toureg wife or had one previously. One was in Niamey, Louis, and one was in Abalak, Bill. Both generally supported such a relationship, citing high morals and ethics and minimal cross-cultural negativity from other expatriates. Another factor for me, believe it or not, was genetics. One of my undergraduate classes was "Livestock Production." In that class, we learned of livestock gestation, livestock ranching strategies, cattle breeds, and cross breeding. The strength of cross breeding two pure breed cattle breeds was something called cross bred vigor. Cross bred vigor was "supposed" to have the strengths of the two pure breeds and minimal negative characteristics from each breed. More generally, to get pure breeds of cows involved a lot of inbreeding (breeding cattle which had the same desirable traits, which often meant that the pairs were often closely related or at least with a similar genetic profile). Inbreeding can allow genetically recessive traits (the genes are present but not expressed physically in the offspring) to be expressed in the offspring. When both of the parents have the same recessive traits, those traits (which are often undesirable) are expressed physically in the offspring. Therefore, my logic was, it was an advantage to marry someone very genetically different than myself. Ultimately, after all the advice I gathered and all the science logic arguments I could dream up, the final decision was just who was the most fun to be with and who had the liveliest character. That was Mariama. This meant cultural differences, possible reservations by the Nigerien civil servants and expatriates who I worked with, and a ready-made family with Mariama and her daughter.

This was a big decision for me, but I was pretty sure I no longer wanted multiple girlfriends. Several dreaded tasks lie ahead. First was to get approval from Mariama's family. I had not let Mariama know my decision or intentions yet. I spoke to Seidi and asked him to go to her mother's house and ask for approval for such a marriage. Seidi headed off on camel. I think Mariama's camp was about 30 km or so to the north. I thought her family would be delighted, but when Seidi returned he said they were skeptical and did not want me stealing their daughter off to America. He argued with them strongly for most of a night, and finally convinced them to agree. There also was the issue of a dowry. The Niger government had set the maximum limit of a dowry at 50,000 CFA (about $150 at the time). I knew I would be in the spotlight, so I insisted that the dowry be no greater than this amount. The next task was to tell Dr. Ali, the project director in Tahoua. When I told him in his office, I sincerely suspect he was completely floored but he had a good poker face. His one requirement was that I had to tell my parents. Two things down, but now I had to call Mom and Dad. I do not recall how I got access to a phone, but I made the call. There was a long pause on the other end of the line. I assured them that Mariama was very special and easy to get along with. I recall that their closing was that the decision was mine, but to consider my decision very carefully.

In the Toureg culture, the wedding ceremony is small, private, and often even secret. That played in my favor as I was trying to keep this on the lowdown as long as I could. So, the Marabou, a religious leader, had come to Seidi's hut and I was supposed to come over late in the evening for the wedding. I was getting antsy as it was getting toward dusk and the wedding time loomed. All of a sudden Dana, my old Peace Corps colleague, walked up to my hut. Where was his truck? Dana had done the same thing I had done going to the previous location of the Peace Corps huts at Ibecetene—he too had taken a run at the same muddy area I had previously gotten stuck in. Dana, though, had picked a narrower part and had almost made it through. His Scout truck was stuck on the northern edge of the muddy low area. Dana wanted to wait until morning but I insisted that we go pull him out that night. I remember Bachir, one of the government civil servants from Tahoua, was with us that night. So, we loaded up two of the Scout pickups with the herders and volunteers and were about to leave camp when Seidi walked up. I asked him if he wanted to help get Dana's truck out. He looked at me with a funny look and said he had a special guest so he could not go. I then realized that guest was the Marabou for my wedding. But there was no frantic indication from Seidi that I should join them right now. I fully expected to be back in 10 minutes or so.

We pulled with both of our Scouts, but could not budge Dana's truck. We adjusted the position of pulling vehicles, as the spinning wheels had dug small depressions. Bachir re-attached his barbwire tow wire as his had come lose. So, 1, 2, 3, PULL and the front bumper mounted grill protector of Dana's truck was pulled off. That was where Bachir had re-attached his tow wire. Dana was livid, asking who had attached the tow wire there and removed the "nose" of his truck. I think it took 2 to 3 hours to finally get Dana's truck out and we all were mostly covered in mud. Seidi came out to hear our stories of getting Dana's truck out. I was expecting the wedding ceremony hand been delayed until I arrived, but Saidi said something to cue me in, that I was not needed. I had missed my own wedding!!

Seidi later told me that it was accepted to have a stand-in for a novice groom at the wedding. Some say it can bring bad luck by pointing too much attention on the groom. Another version was that they cut the groom a little slack if he is the shy type the first go marriage.

In the morning by the time I got up, Marabou's camel was gone, so I knew he had left. I casually asked Seidi were Mariama was and he said she had left with her brother early in the morning by camel for her mother's house. This must be to prevent any hanky-panky until the more public wedding reception is complete and the new bride comes to the groom's house.

The task was to now prepare for the wedding reception. The reception, as I recall, was three weeks to a month after the wedding. It gave the bride's family time to prepare some gifts to the groom (at least that is what I was told). The groom has a heavy lift. He is to prepare a reception with food, tea (always), tobacco (many Toureg women chewed tobacco), and ornate clothing for the new bride. I announced our marriage and reception at the ranch headquarters, to the volunteers at both government ranches, and some of the staff at the project headquarters in Tahoua. I had a tailor, who made some shirts for me at the Tahoua market. I asked him about making a fancy set of clothes for my new Toureg bride and he said that he knew how to do that very well. While I was at the Dakoro Ranch weighing the cattle that month, Seidi had told me that I should buy the tobacco there, as some of the best was available there. It was on my trip back from getting supplies in Maradi that I took advantage of the Dakoro market day. After some searching and asking several vendors, I found the tobacco section. I selected a vendor who seemed to have a wide selection and looked friendly and, hopefully, honest. I knew I had virtually no chance of getting a fair deal, being the only white, and assumedly rich and naive, person in the market. My approach was to just be blatantly honest with the guy. I told him I needed some of his top-quality tobacco and was told that the Dakoro market was the place to get it. He dug around and pulled out a bundle of tobacco leaves tied with a date palm leaf strip. The bundle was very aromatic (nearly causing me to start coughing) and was about the size of four large flip-flop sandals tied together. I told the vendor that I wanted him to give me a reasonable price and informed him that I had no idea what that was. He said he would give me a fair price and threw out a price. I asked him again, if he was sure this was top quality tobacco. Then I asked him to confirm that his price was reasonable. He replied "yes" to both.

There always a group of folk following me, the foreigner, around through whatever market I was in, like I was a zoo animal or something. Often it was kids, but today it was several 19-year-old-ish young males. I think they were somewhat astounded by me speaking Housa as well as I did. I watched their facial expressions during the negotiations with the tobacco vendor and there was no indication in their faces or body language that I was being taken to the cleaners. So, I gave the vendor the money and said that it better be good tobacco or I would come back next market day to complain to him. He said confidently a third time that this was very good tobacco.

I had paid several of our herders at Ibecetene to prepare the meal, providing meat, rice, etc. for the reception. The day before the reception, Seidi was going over the list of all the stuff he needed to plead with Mariama's family to let her come to my house. This meant large amounts of tea, sugar, and tobacco. Seidi also asked to see the new bride's set of new clothing. I dug them out from behind my truck's seat. He looked them over and said they were fine. Seidi wandered off to do something else. I noticed the bride's clothes sitting on my truck's hood, so I grabbed them and put them back behind the seat in my truck. As I remember, it was a project driver who took Seidi to my wife's house.

Everything was in motion and I thought I had prepared well. I went to bed that night, not expecting Seidi to bring Mariama back until mid-morning or so the next day, the day of the reception. I was sleeping out by my truck on my sleeping mat in front of a mud walled thatched hut which was for me. I was awakened about 2 a.m. by a revving engine and a pair of glaring headlights. I got up pulling on my pants as I squinted trying to see who had arrived. The thought had occurred to me that, maybe there was some objection to the wedding from the Préfet's (equivalent to a state governor) office in Tahoua or something. I was relieved to see both Mariama and Seidi! Seidi came quickly up to me and said, "Where are the bride's clothes?" I said they are in my truck and promptly produced them. The clothing was quickly handed over to Mariama for her inspection.

Seidi explained that the family almost did not allow Mariama to leave their camp. You see, the bride is supposed to come to the groom's house all dressed up in fancy clothes. I noticed Mariama's hair was all done up in small braids arranged in an intricate design. Her hands were also decorated with henna. Seidi said, and the driver confirmed, that he had negotiated very hard and long to get the approval from the family. They only agreed if Mariama could come in the dead of night when no one was around, so she could have the fancy clothes on when folks saw her the day of the reception.

Seidi and the driver headed off, leaving the two of us alone together. It was a bit awkward finally being together and alone after being separated for a month. We spent our first night together out under the stars on my sleeping mat. The next morning Mariama was up early. As I recall I was not supposed to see the bride much until after the reception (kind of irrelevant since we just spent the night together, but few knew those details), so Mariama was soon surrounded by a flock of women all talking at once and drinking tea. I hung out with the men outside on a different side of my hut.

There was some conversation amongst us men, but once we stopped just to listen to the drone of conversation coming from the women's side of my hut. It was a constant drone of noise, with no chance of deciphering what any single woman was saying.

At a wedding reception, one or two folks are supposed to be jokesters to tease the groom. I had been warned and there was one guy constantly teasing me, constantly asking for a gift of tea and sugar. I was bothered and not sure how I should respond, then it came to me, I could just go ask my wife! This was probably breaking all the remaining protocols I had not already broken, but I just walked over to the women's group and asked where Mariama was, stating that I needed to talk to her. The women pointed Mariama out. (The women cover their heads, particularly at social gatherings, so it's not easy to identify a particular woman, no matter how well you know her.) I whispered a question about how I should respond to the jester guy. She just smiled and calmly said, "Just give him some tea and sugar." So, I gabbed some tea and sugar out of my cache in my truck, gave it to the jokester, and he promptly laid off the harassment.

So, it's true that I was not at my own wedding. About 30 years later I asked Mariama about the details of the wedding I had missed. She said she could not see anything as she was hiding behind a grass windscreen so the Marabou could not see her. From her description, it sounded like the whole affair was over in 15 minutes or less. A couple of months later, Mariama and I were talking about our future and she announced that she did not expect our wedding to last more than a couple of years. This seemed reasonable to me, as I was having trouble seeing or forcing her to adapt to life in the U.S. Our relationship continued to expand and produced two children. First, I started taking Mariama to Tahoua with me, and later we lived in Tahoua, and finally we moved to the capital city of Niamey. Ultimately, we moved to the United States. Now we have been married for over 35 years. That is how you eat an elephant, one bite at a time.

— 17 —

HONEYMOON (1983)

Mariama and I had been married for more than a year before we were able to go on our honeymoon. This delay was largely caused by my busy work schedule. In the summer and fall, I had a lot of work-related field campaigns and reports to get completed. Also, I had to get all the necessary papers for her to get a passport. Mariama and her now two-year-old daughter, Aminata, were on the same passport. I was strongly encouraged by several friends in Niamey to take Mariama on a honeymoon trip, despite it being a completely new concept to her. One of our friends in Niamey was Peter, who had been a Peace Corps volunteer in Togo. Peter advocated we go there for our honeymoon and gave some detailed destinations we should see while in Togo. I took detailed notes as these destinations sounded very promising.

We had come down to Niamey to get the visas for Togo and make the flight reservations. Peter let us stay at his place in Niamey. I thought I had all the necessary paperwork ready. We went to the airport, checked our bags, but then in the final boarding check Mariama was denied access to the plane. Apparently, there was an additional approval form that needed signatures before she could leave the country. This sounded suspicious to me. I thought that was what the passport and visa were for. As we were out on the tarmac with the police officer to reclaim our baggage off the plane, I asked the police officer what was the deal with this form as I had never heard of it. The policeman said it was no big deal and routine. He also said the form was fairly easy to obtain.

After a couple of days, we had the signed form allowing Mariama and Aminata to leave the country, and I had changed our flight reservations. This time we made it on the plane. It was a dual propeller plane with about 30 passengers. Mariama wanted to sit by the window. This was her first time flying, but she seemed remarkably calm. Takeoff was routine with a nice view of Niamey. Soon we were at cruising altitude and as I recall the flight was about an hour or so in length. There were small, scattered popcorn clouds on a mostly clear, nearly windless day. Mariama tapped my shoulder and wanted me to warn the pilot to watch out for the clouds as we were getting close to some of them. She did not want the pilot to hit any clouds. Before I could respond, we had flown through a cloud. I explained that clouds were like smoke and were not solid.

Once on the ground we found ourselves in a big city, the capital of Lomé. We found a hotel room and went for a walk. Mariama wanted her hair braided in corn rows as she observed on many of the women on the street. We came across a hair dresser shop and they agreed. They said I should come back in an hour or so, and I needed to take Aminata with me. Aminata and I knew each other, but we had never been together alone without Mariama or an uncle or aunt of Mariama's in attendance. We started off OK and we walked off down the sidewalk as I constantly tried to distract her with questions in Housa, even though I was not sure Aminata even understood Housa. Aminata soon noticed her mother was not there and started crying and nothing I could say or do would even dim the volume of her crying a fraction. I had felt conspicuous earlier in the walk, not being the only white person, but a white person who appeared to be kidnapping a young girl against her will. People on the street were giving us weird looks. I finally just gave up, as I felt that by trying not to look like a kidnapper, I was really only looking more just like one! Finally, we rounded a corner and were within sight of the hairstylist. We burst through the front door of the hair shop with Aminata in full wail and tears streaming down her face. Mariama just smiled and said something to Aminata that calmed her considerably. There were 4 or 5 gals all braiding away on Mariama's long black hair which extended to the middle of her back. The hair stylist crew agreed Aminata could stay as they were almost done. I waited outside, trying unsuccessfully to look normal. Mariama and Aminata finally emerged from the hair shop. Once we got out of earshot (no one there spoke Housa, so I am not sure we needed to even do that) Mariama said they had braided her hair too tight and it was hurting her head. Mariama left the braids in for a day before she took them out. When she did, she found that the hairstylists had actually broken some of her hair at a length of about a foot long. So, when Mariama put her hair up in a looser couple of big braids, there where these stray hairs about a foot log sticking up on the top of her head. I said she looked like a rooster—fortunately, she laughed.

One thing that struck both Mariama and me was that the goats in the country side of Togo were pygmy goats about one-third the size of what we knew in Niger and what I knew from the States. This alarmed Mariama, as when we ate at a restaurant she was always suspect about what kind of meat she was eating. She did not want to be tricked into eating pig or dog or something else. Given these small-statured goats, she was unsure she could identify goat by the bones or not. At one restaurant, I scanned the menu and mentioned to Mariama that they had horse steaks on the menu. To me it was a mere curiosity but to Mariama is was an abomination. She demanded we leave immediately, and we did, despite my arguments and pleading. After that, Mariama only ate chicken. I pointed out that was probably why she had that rooster waddle of hair on her head. As I recall, she only found that mildly humorous (no matter how many times I used it). In retrospect, I am probably just lucky she let me live.

Our friend Peter back in Niamey had recommended one specific beach for us to go swimming. Peter had said that public beaches were often used as outhouses and recommended this one remote beach as a clean beach. I thought it would be interesting for Mariama and Aminata to play in the waves and witness the salty water. The taxi driver dropped us off in a small parking lot with one car in it. There were palm trees between us and the beach, but I could see and hear the ocean 40 feet or so in front of us.

I asked the taxi driver how I was supposed to get a taxi back to our hotel and he assured me that the highway we had just turned off of into the parking lot had lots of taxi traffic. So, I confidently led Aminata and Mariama onto the beach. There sat a young Caucasian girl sunbathing with no top. I strongly suspected Peter had sent us to a topless beach, and I decided to just act like this was a normal thing for white folk. Many of the pastoral women in the bush back in Niger do not hesitate to breast feed in front of family members and family friends. But if you are a stranger, some of the Toureg women would try and hide out of sight or even wrap up in a mat (palm reed or plastic reed) around them. So, I was hoping Mariama would not think this too peculiar and Aminata was just too young to know or care. We were there about 20 minutes before the Caucasian woman picked up her towel briskly and walked back to her car in an apparent huff. She promptly drove off. I guess she did not like the idea of the topless beach being a family beach. I was startled for one second but realized that I was not interested in the Caucasian gal's approval AND we now had the beach to ourselves. Mariama, however, was nervous about Aminata playing in the waves. She was worried her daughter could be sucked out to sea. Mariama had also picked up on the Caucasian gal's anger and suggested that this may not be a family beach since there were no other families here. After 30–40 minutes we were back at the highway hailing a cab back to the hotel.

A second destination suggested by Peter was a hike to a waterfall out in the rain forest. This seemed to be an interesting and safe bet. I followed Peter's directions on which bush taxis to get and which lodge (collection of large grass roofed huts) to ask for a guide to the waterfall (along with the approximate correct price to pay the guide). Soon a 16-year-old or so lanky boy agreed to guide us to the waterfall. We wove our way through the forest and saw bananas. However, our guide said that they belonged to someone and we could not eat any. The trail was barely a game trail in many sections so without the guide, we would have been hopelessly lost. We continued deeper into the forest where the trees were bigger and taller and the underbrush thicker. The trail became steeper and rockier. The final section (this was about an hour in) had a fallen tree across the path traversing a fairly steep side slope. Our guide asked me to let him carry Aminata through this last rough section, and I agreed. We proceeded carefully and finally came to the waterfall. It was about 20 feet tall and the creek coming out below the falls was less than two feet across. It was very scenic and well worth the time and effort. Mariama enjoyed it, though initially, she did not trust our guide.

The final recommended destination was a lake with no snails which can carry a parasitic worm which can cause people to go blind (river blindness) and where we could wind surf. Again, Peter's directions were spot on. When Mariama saw what a wind surf was, there was no way she was going to do that. Many Touregs do not know how to swim. Being a desert and Sahel ethnic group, swimming is not a traditional skill. In the dry season may of the intermediate ponds dry up. Shallow (6 to 15 feet) wells are then dug so drinking water and water for livestock can be provided for 2-4 months. When the rains return, the ponds refill. Kids play and swim in these ponds in the hot and humid rainy season. If a kid steps into one of those old wells, which are only 6 feet in diameter at the most, they cannot swim the 2–3 feet to get to shallow water. A fair number of kids and even some adults lose their lives that way. I suspect that is one of the reasons for Mariama's fear of water, but now she can swim a bit, up to 10 feet or so, mostly underwater.

I rented a wind surf and got some verbal instructions and a quick demonstration from the fellow renting the wind surfs. I fell a lot. After about 30 minutes, I could go straight but had not figured out how to turn. I wind surfed to the middle of the moderate-sized shallow lake, stopped, fell, turned the board around, and then wind surfed back. The board renter wanted me to pay him for lessons, but I was too frugal. I was tired, too, and tired of having Mariama laugh at me. She enjoyed watching me fall.

One day in the market, we noticed an older Toureg guy following us. We stopped to talk with him. Ironically, I was the only one who could talk with him. He had been there over 10 years and had completely forgotten Tomacheque. This stunned both Mariama and myself. After 10 minutes of questions and responses, we gave him some money and parted ways.

We returned to Niamey with no major incidents. But once in Niamey, Peter wanted a report from us on how we enjoyed the trip.

18

IT'S A BOY (1983-1985)

We were on our honeymoon in Togo when Mariama got pregnant. We did not know for sure though until we got back to our house in Tahoua. Mariama had me take her to a French female doctor friend of ours, Domonique, who confirmed the pregnancy. Within a couple of months, phase three of the Niger Range and Livestock was beginning. I was told by one of the Tufts University team that in order to justify my being on the project, I needed to have a skill set that none of the Nigerien civil servants had. He recommended that I return to New Mexico State University, a sub-contractor on the project, and take statistics and computer programming. He saw the skill set of a qualitative range specialist as an asset to the project. I had done poorly and did not like statistics in my undergraduate class. As part of that class we had to do some assignments by logging on to the campus computer system. I logged on and hated it. I think that was one of only two classes where I got a grade as low as a C. I thought I had submitted the computer homework on the computer, but apparently, I had not submitted it correctly and that was what I suspect was the cause of the C grade. So, I was kind of sour on both computers and statistics, but I had no choice as I wanted to continue working in Niger on the project.

Mariama, Aminata, and I returned home to Mom and Dad's place, which was now a small town in northeastern North Dakota, 40 miles from the Canadian border. It was early summer, and I had bought a car from my older brother while I was still in Niger. It was a four-cylinder compact two-door Ford Mustang. Mom and Dad had moved most of my stuff up from Montana, and I selected a subset of useful stuff for our trip down to Las Cruces, New Mexico. I had a bike rack on the trunk door with a ten-speed bike I had bought from Andy, my roommate during my undergraduate years at the University of Montana. I forget what route we took down to Las Cruces, but we passed through Grand Junction, Colorado, Alamogordo, New Mexico, and White Sands National Monument. Mariama really liked White Sands.

When we arrived in Las Cruces, I looked up a potential roommate contact that my advisor had provided to me, a graduate student in the Range Department. The house and bedroom looked quite acceptable to me and it was close to campus. The trouble was that the graduate student did not want to have a family as a roommate. Mariama and I were unclear how we were going to do this college graduate degree. I was kind of letting her make up her own mind if she was going to stay with me in Las Cruces or return to Niger. I made her aware that I would have to study hard to be able to do well in tough classes like graduate level statistics and computer programming, which my advisor had me taking. Ultimately, Mariama decided to return to Niger with Aminata. It was good and bad news for me.

I sent letters and recorded letters on audio cassettes, which I sent to Mariama in the mail. I only got a few short letter responses from some of the expatriates at the project who Seidi, her uncle, kind of knew. I was able to wire money to my bank account in Tahoua, where Mariama or Seidi could pick it up. I buried myself in my classes and did well in three different statistics courses and two computer programing classes over a year and a half. My advisor had me take some desert ecology and landscape ecology classes as well. I had been accepted at New Mexico State University on probation, as my GRE test scores were marginal (but I had just been in a third world country for 4 plus years). But after getting all As and Bs my first semester, my advisor said I was off probation.

I had a few friends that were fellow graduate students in the Range Department, but I mostly filled my weekends by mountain biking in the desert around Las Cruces. I would just pick a compass bearing and head off through the desert bushwhacking. There was only one close encounter with a rattlesnake. I had just cranked up a two-rutted jeep trail in the Dana Mountains just north of town, when I heard a psssss. Nuts, I thought I must have a flat tire! I was going real slow, but rather than apply the brakes, I let the bike roll an extra yard or so to a stop. I jumped off and stuck my head down by the crank as this psssss sounded like it had come from between my heels when I was on the bike. But both tires were firm with air and there was no hissing sound. Confused, I took a step back, scratching my head and looked back on the road where I had heard the sound. To my horror, what I thought was a dried cow paddy uncoiled and a rattlesnake about a yard long slithered off the road. Needless to say, I was nervous mountain biking the rest of the day around the creosote and mesquite bushes.

A couple of weeks after New Year's Day, I got a letter that I had a son, Abdoul, who was born on New Year's Day. Also, mother and baby were well. And that was about it.

After what was about a year and half, my advisor informed me that they wanted me to return to Niger and begin work. He said that I had the skills needed for the job but would still need to finish my master's thesis later. So off to Niger I went and met the head Range consultant there, also from the Las Cruces area. We had met briefly before he headed over to Niger. Many of the same Nigerien civil servants I knew before were still working on the project. I spent most of my time in the computer room working on portable computers the size of moderate suitcases. The keyboard would latch over the front of the monitor to form a suitcase like thing with a handle. The monitor was lime green monochrome.

I was particularly interested in seeing my son. I had talked to the Peace Corps volunteer who worked up in Kao, a small town near the Ibecetene Ranch, who had seen my wife and son. The volunteer was working as a nutritionist for babies. She was visiting Tahoua at a party I went to and I tracked her down. She said she had seen my son and that he looked like all the other Toureg children and that he was healthy. She did not seem overly interested in my situation. I awkwardly and quickly moved on to another group for conversation.

I had asked the project director—not sure if it was Dr. Badamasi who replaced Dr. Ali—but after working about two weeks, they gave me a driver and a Toyota pickup to go get Mariama and the kids. I was thankful for the vehicle, as Dr. Badamasi was strict about expatriates going to the bush without a driver and usually only for work-related issues. We found Mariama's family's camp by just asking folk we saw on the road or later folk out in the bush as we traversed off road. Most Touregs seemed to know Seidi or, if not, they surely knew his mother and Mariama's grandmother, Fatu. Fatu was a famous traditional violinist and had many recordings of herself accompanying various Toureg singers (always male). I was happy to see familiar faces, but nervous about not breaking any traditions and customs. As folks were helping Mariama pack, someone dropped my son on the mat next to me. He was getting close to one year old and could sit up wobbly on his own and stared at me curiously, but unafraid. I must admit, I felt a little conspicuous being evaluated by my son. I should have played with him, but we kind of just quietly checked each other out. I was afraid if I said anything or reached out toward him, he would start crying.

Having learned a bit from my past mistakes, I came bearing gifts of tea, sugar, tobacco, and most importantly some new clothes for Mariama! I did not want to hurry her or her family, but the driver and I had to be back in Tahoua that night. Everyone seemed to understand the importance of our time frame and as dusk was falling, we climbed into the cab of the truck and headed to Tahoua. In the dark of the cab I put my arm around Mariama and she snuggled closer. We assumed the driver could not see us snuggling, but I think we both knew he was watching us out of the corner of his eyes in the cab, dimly lit by the dashboard lights. So again, in the dead of night, my bride returns to my house, but this time it is to a modern cement brick house with an air conditioner, two bathrooms, a refrigerator, and a chest freezer.

Mariama's family—it seemed that I was now related to half of the Arrondissiment (aka county)—would stop by our house when traveling through or visiting the Tahoua market. We would have 7–10 camels tied up in our concession with what seemed like 15–20 guests on weekends and market day. We bought a little 4-wheel drive Suzuki jeep-like hard top. With that we could go visit her family as needed.

So, I was not at my wedding and also was not there for my son's birth. I was batting a thousand.

— 19 —

GROUND TRUTHING (1986-1991)

With Phase 3 of the livestock project in Niger, the emphasis of the range management section was transitioning to a pasture forage mapping effort. This was because the Sahel has variable rainfall from year to year. We have all heard of the droughts in West Africa, food aid, and human health support efforts. The grasses in the Sahel are dominated by annual grasses, which means they grow from seed every year and die at the end of the rainy season. Pastoralists move where the grass is with their livestock. So, in a drought, the price of livestock crashes to next to nothing while the price of cereal grains for human consumption spikes. This puts folks who are one hundred percent dependent on livestock for their livelihood at an extreme economic disadvantage. Droughts have brutal impacts on farmers, too, but at least the commodity they are selling still has value or even increased value. Thus, the livestock forage mapping effort was to inform government officials and herders of places with grass and estimates of the amount of grass. This also would allow food aid organizations to know of impending drought impacts.

The project was working with ILCA, the International Livestock Center for Africa, to bring in an airplane-based monitoring coupled with a satellite monitoring effort. Both of these monitoring approaches needed estimates of pasture production levels at different locations across the pastoral zone of Niger (the northern two-thirds of the country where it was just too dry to farm). Our work was to visit selected locations on the ground and measure the forage production. We were to sample 30 or so locations within 10 or so days late in the rainy season. Originally, the area we sampled on the ground was small (less than 1 km^2 area). Our sample plots were then then shot with a radiometer on a tripod to capture the amount of light reflecting in different wavelengths from the ground and vegetation within our plot. The airplane then collected the same reflectance information over the area where our plots were and then over the surrounding 10 by 10 km area. This was so that the forage production estimates from our ground could be scaled up to a larger area that could be more reliably linked to satellite data, which was a fairly coarse resolution (1 km). The satellite digital images could then be converted to pasture production maps. The first year, our job was designing our ground sampling strategy, training our crews, and getting the equipment and vehicles we would need. This worked reasonably well the first year we did it. ILCA produced a report, a pasture production map, and gave a training session on remote sensing at our office in the dry season.

In the second year, however, the contract with ILCA fell through. I was informed that ILCA and their airplane and satellite experts would not be coming. It was already mid-rainy season when I learned this. I was devastated. Being a problem solver, I asked myself if there was anything we could do. I came up with an idea of extending our ground sampling across the 10 x 10 km area at each of the 30 or so locations (ground truth sites). My plan was to have ground crews drive in random zig-zag lines traversing the 10 x 10 km block, sampling the grass at random stopping distances. This would provide an estimate of the pasture production for the 10 x 10 km areas, which could then be correlated with the satellite data. I presented this to the project director, Mai Daji, and he was stunned. He asked me to confirm his understanding of what I had just proposed. He had correctly understood my idea and liked it. I assumed he would run it "up the ladder" to get approvals from both the Niger government and USAID. We got the go ahead and I called my remote sensing professor, John Harrington, back at New Mexico State University, and he agreed to get the satellite data we needed.

Within a week we had two crews off sampling the ground truth sites with my method. I was curious if it would work or not. How long would it take to sample a site? Would we be constantly getting stuck with our vehicles? All worked reasonably well, but working in the rainy season we did occasionally get stuck out in the middle of nowhere. We were always somehow able to get our vehicles unstuck. A fallback was that there were often livestock herders usually in the vicinity with their livestock, so we could have tried to get help from them. One real "head scratcher" of one particular "being stuck" incident was that there was no real obvious reason why we were stuck. No big mud, sand, hole, or anything! Our crew of four were all stumped. I had the driver try to drive forward and spin the wheels as I walked around the vehicle. Then I told him he was not in four-wheel drive as neither of the front tires were spinning when he tried to move forward. He showed me that the four-wheel drive was indeed engaged in the cab. I was standing by one for the front tires completely confused when I noticed the front four-wheel hub locks was wiped clean as if it had recently been adjusted. On closer inspection, sure enough, the hub was not engaged. In disbelief, I called the driver over to see this. He looked back along our tracks and spotted a small stout curved stick that had popped up when we drove over it and had disengaged the front hub. We were relieved but also a bit mad that we had wasted so much time. The driver, in a half angry mood, reached down and engaged the hub, yelling to everyone to climb in! Everyone piled back in the vehicle, as we were all confident we had solved the problem, and we had!

After our field data collection field trips of 10 days, we would crunch the numbers and get our pasture production estimates at all the ground truth sites. These pasture production estimates where sent to my professor at New Mexico State University. John Harrington correlated our pasture estimates with the satellite data and produced color-coded maps. We had a big workshop down in Niamey shortly after. John had sent the color pasture maps via DHL to me in Niamey. It was the second day of the workshop. We had gone over our ground sampling protocols and our ground truth site estimates. I could stall no longer, I had nothing more to say, when in walked the project director, Mai Daji, with the color pasture production maps. Talk about timing! (see http://pdf.usaid.gov/pdf_docs/PNABG559.pdf for more details)

We continued this sampling approach for 4 or 5 years, doing mid-rainy season and late rainy season maps and a report. It was a lot of time in the bush sampling across the country. In the later years, they developed a satellite receiving station in Niamey, and I was able to get the satellite data there as well as remote sensing software and a big computer tape reader needed for getting the satellite data onto the computer.

— 20 —

SKINNY (1986)

After several years of living in Tahoua, Mariama and I decided to invest in some property in Tahoua. We bought a residential lot on the north side of town. I designed an adobe house that could be rented to two families. We hired a man to build it. The doors and windows were metal and hinged to open. The roof had wood beams covered by old barrels split open and then covered in mud. The roof only leaked a couple of times and the guy who built it was five houses down the street and had agreed to fix the roof anytime it leaked. The bathroom was an outhouse in the corner of our lot. We later added running water with an outside spigot, an indoor shower (no hot water), and flush toilet. No electricity!

After completion, I asked the project if it was OK for us to move into our adobe home. The project would no longer have to pay the rent on the fancy modern cement house. The project agreed. Seidi was our backdoor neighbor, and we built a gate to facilitate moving between our two mud walled concessions.

At this time, I was working hard at the project doing computer programing, analyzing field data, and organizing and participating in our ground truth field trips. Mariama was making an effort to cook, and she even hired a woman to help her. Mariama was a terrible cook—well, when there were guests she did very well, it was just in the daily routine that she showed little creativity. It was rice and tomato paste or some kind of pasta day in and day out. I just lost my appetite. I would get hungry at night after going to bed, but would just force myself to suck it up. Mariama sometimes cooked beans, and occasionally we had milk. All of our kids and Mariama all had healthy weights. I often skipped breakfast or only had coffee. In high school, I was 5'11", which means an ideal weight should have been around 156 lbs., with a lower healthy limit of 139 lbs. At this time, my weight was 130 lbs. and I was a walking skeleton. Folks at the office and Mariama's extended family members all said I was too skinny. Even the expatriate doctor on the project said I was too skinny.

One of the project expatriate veterinarians on the project had lost a lot of weight when he was sick in Tahoua for 3–4 days. He advocated carrying a little extra weight as a buffer against such illnesses. At 130 lbs., I had no or very little reserves if I was sick and unable to keep food down.

I started buying grilled meat at the market and bringing vegetables home, but weight gain was slow. I also started eating something in the morning. Gradually, I was able to put some weight on.

--- 21 ---

IT'S A GIRL (1986)

Mariama simply stated one day that she needed to go see the woman's doctor and our French friend, Dominique. I took her in our Suzuki SUV to the women's and young children's clinic. We waited in line with all the other women and their kids until it was our turn. Mariama wanted to go in alone, so I waited outside. I was mildly curious but doubted it was anything serious, as Mariama seemed perfectly normal. Within 5 minutes, Dominique opened the door and called me in. She informed me that Mariama was pregnant. This was mind blowing to me as she was still lactating, feeding our one-year-old son. We thought that lactating meant a lower probability of Mariama getting pregnant. I had also been applying the rhythm method, to further reduce the probability of Mariama getting pregnant. I complained to Dominique that this was not planned. I said that Mariama and I needed to discuss this, as I was completely blindsided.

After a couple of days and discussions with Mariama, I was coming to terms with another child on the way. We proceeded with the pre-birth programs in Niger, and with routine visits to see Dominique. Mariama wanted another son to play with Abdoul, but I kind of preferred a daughter. In Toureg culture, the sons typically care for and provide for their aging mothers. Daughters are whisked away to the groom's family and often have little or no resources. Also, mothers tend to re-focus their priorities on their own children's needs. In the final stages of pregnancy, Dominique announced that Mariama's baby was in the breech position. Try as she might,

Dominique could not rotate the child in Mariama's womb. Dominique recommended that we either go to a missionary clinic south of Tahoua or go to Niamey. She said the Egyptian surgeon at the Tahoua hospital had lost the lives of the last five C-sections he had done.

Mariama had paid her family Marabou to do some prayers to help ensure a good birth and for a healthy baby. Her confidence was so strong in this that she wanted to stay in Tahoua for the birth. I argued strongly against that. I let Mariama know that I had confidence in Dominique's knowledge as well as her knowledge of the history and skills of the Tahoua physicians. One day Mariama and I were standing in line for a routine examination by Dominique, when I turned around and saw Mariama's family Marabou in line a couple of people behind us. I smiled and nodded and he smiled back. I turned toward Mariama, who was in front of me in line, and whispered in her ear, "If that Marabou is so confident about his prayers, why is he in line to see Dominique?" Mariama sighed and said, "OK, we will go to where there is a better doctor!"

I wrote a letter to the missionary clinic explaining our predicament and Dominque's recommendation. The response letter declined our request, saying we were not in the target populations they were focused on. This seemed odd and counterintuitive to me at the time, but later I guess I can see why they would not want to become a hospital for expatriates. It still seemed odd to be turned away by missionaries in a time of need. Dominique recommended a clinic in Niamey and sent Mariama's health records there by mail. Dominique recommended we go down to Niamey a month early. This was possible as we knew a couple in Niamey, David and Karen, who had been the previous range supervisor on our project when I was in the Peace Corps. Karen was a volunteer the same time as me. I informed the project director in Tahoua, and he approved my trip to Niamey for a month. Karen and David had two young girls, and they played with our son Abdoul and daughter Aminata. I tried hard to help out with chores around the house as much as I could. I am not an experienced or a good cook, so mostly I washed dishes and did other tasks.

Mariama and I did preliminary visits at the Niamey clinic. The clinic staff were well-versed in Mariama's condition thanks to Dominique. We were approved for delivery when labor began. It was mostly just sit and wait and try to help out around the house as much as I could. David and Karen had a nanny to help with their kids. Mariama and the nanny kind of became friends. After about 3 weeks, Mariama stated that labor had started. I grabbed Mariama's prepared "go" bag and we were off to the clinic. They took her in for examination within five minutes of arriving. I waited anxiously in the waiting room for maybe 20 minutes. Then they called me in and told me Mariama only had indigestion. I was stunned that the mother of two could be tricked by indigestion. We returned to David and Karen's house, rather embarrassed.

About a week later, the nanny came to me while I was out watching the kids play. She said that Mariama was in labor and it was time to take her to the clinic. Mariama had only told me she had not slept well and wanted to sleep in that morning. Off we went yet again to the clinic with me wondering if this would be another false alarm. We were informed in the waiting room that the dictator of Niger's (Seyni Kountché) daughter was already giving birth in their sole birthing room. They had Mariama go into one of their rooms where they inspected her and returned her to the waiting room. So, I was sitting next to my wife in the waiting room half expecting her water to break any time. Everyone else in the waiting room seemed concerned about Mariama as well, but now I wonder if they were the other mother-to-be's family. Mariama sat quietly but sometimes wiggled and winced. I think we must have waited 40 minutes to 1 hour in the waiting room before we were invited into the birthing room. I thought I was invited in only to translate. After talking with the French midwife, I learned that she only knew a little D'jerma, which Mariama did not speak. I thought by teaching the French midwife "push" in Housa, I could then leave. But that never happened so I just stayed. They tried to rotate the baby in the womb, but that did not work and really put Mariama in a lot of pain. So, they, the midwife, and a doctor, were going to do a breech delivery. Breech—butt first—is more difficult for the mother and more dangerous.

I was relying on their expertise and assumed that the doctor was there for C-section surgery if needed. There was a lot of yelling "push," and I would encourage Mariama to breathe as I had heard that mitigated labor pain. I would hug Mariama when she relaxed between contractions and hold her hand during contractions. Apparently, progress was not going well, as the doctor started nearly laying over the top of Mariama, pushing the baby's head by pressing on the top of Mariama's stomach. There was lots of loud yelling at this point, like it was a football game or something.

Finally, the baby's legs appeared and the midwife announced a daughter. And then the head appeared. The baby was not breathing and the midwife quickly laid the baby on the table and started mouth-to-mouth. Soon the baby was breathing and started crying. The midwife re-set the dislocated hips. The midwife pushed Mariama's stomach with her straightened fingers as if trying to penetrate her stomach. The afterbirth immediately ejected. They wanted Mariama to breast feed the baby, but initially she refused. She had just undergone excruciating pain and was completely shot. I explained that it would help the baby fight disease. Mariama finally nursed the baby for 3 minutes.

Mariama was put in a room with a bed for recovery. She was really tired and just wanted to rest. The nurse asked if she wanted the baby in her room and she said "No." Mariama said I should go back and let our kids and Karen and David know the result and that both the mother and baby were well. It now was late afternoon, and the clinic had limited visiting hours, so I said I would return the next morning. I think Mariama had me contact a Toureg friend of hers to let them know as well. After I left, Mariama pressed the nurse call button and had the baby moved into her room.

The next morning, Mariama was still exhausted, but the baby girl had stolen Mariama's heart within 5 hours after the birth. Before noon, the nurse announced that the Marabou was there and we said to show him in. The Marabou was there to give the baby her name. I suggested several names, but the name was assigned based on the day of the week the child was born, so she was named Rai. I pointed out that her mother had a young daughter with the same name, but apparently that was irrelevant. Later that afternoon a young kid, 20-years-old, showed up with a tablet and a pen. He was there to make the birth certificate. He had the day of birth from the front desk, but he got our names, the baby's name etc. He tore out the original and handed it to us. He must have kept the carbon copy.

As I recall, Mariama was allowed to check out midday on the second day after the birth. The midwife came to our room and advocated that Mariama have liver several times and eat nutritious food in the next couple of weeks. The midwife also said that the outcome would probably not have been as bright if Mariama had tried to give birth in the bush. I brought clothes and a baby basket Mariama had bought earlier. We returned to David and Karen's house for a couple of days and took one day to visit a Toureg livestock civil servant agent we knew who was working 60 km outside of Niamey. Mariama enjoyed visiting with him and his wives, he had two at the time, which is extremely rare as Toureg women usually do not put up with a second or more wife. I took some pictures there and when they were printed, we could see how fatigued Mariama was. Soon we returned to Tahoua. Mariama and the baby rode in the project's Peugeot station wagon with some of the project staff while I took Aminata and Abdoul home in our Suzuki SUV. She got to our house a good hour or two before we did. Both Mariama's mother and grandmother were waiting at our house to see and help care for the new baby.

In retrospect, there were lots of decisions or forks in the road during this adventure. It seems that Mariama and I navigated all of those forks well enough to escape with our new, healthy daughter. But I was shocked about the pain and impact the birth had on my wife. Given that, my personal goal to restrict my biological children to two was strengthened and reinforced. I wanted only to replicate my DNA (two times 50% equals 100%), minimally contribute to overpopulation of our earth, and further avoid the pain of childbirth on my wife.

— 22 —

AIR MOUNTAINS (1986)

About a week after our daughter Rai's birth, several expatriates on the project invited me and Mariama to go with them on a guided trip into the Air Mountains, north of Agadez, in northern Niger. The Air Mountains are an area famous for rocky desert mountains, history, sand dunes, petroglyphs, and oasis gardens. I needed to check with Mariama, as our new daughter was only a little over two weeks old. This shows how unfamiliar I was with raising a baby, yet I was the father of three. Mariama categorically said she could not go because of our young daughter, but because of her family being camped only 15 km away and her mother and grandmother helping her at our house, I could go. This was a predicament for me. I had always loved visiting Agadez on project field trips. I had been jealous of two Peace Corps volunteers working up north of Agadez in the town of Iferouane in the Air Mountains on a wildlife project. I had heard stories of the oasis gardens and even villages up on top of the Air Mountain plateau, which peaked my curiosity. I had to weigh that against being a dead-beat Dad and deserting my wife in this important life stage. Remember, I missed my wedding and my son's birth! The scales of my decision process were tilting toward going on the Agadez trip, but it became a conclusion when Mariama reassured me that she would be fine and content without me.

There was a Dutch couple from Tahoua and two American expatriate couples from the project, one an economist, Jeff, and his new bride, and Greg, who was on the range management team, and his wife. They had done all the arrangements and I suspect they invited me because of my local language skills, familiarity with bush camping, and to help defray costs. Plus, in my humble opinion, I think I am easy to get along with and fun to be around.

I think we drove up in our personal cars, as nearly everyone on the project had different versions of the little Suzuki SUVs. I must have squeezed in one of their vehicles. Once at Agadez, we got hotel rooms, and the next day we met our guide for the trip. He was a tall, slender light complexioned Toureg, who I liked instantly. Initially, I said little as our guide spoke French, which all of my expatriate comrades spoke. He had one of the two vehicles that we would be riding in, an older Toyota SUV. He confirmed that we would not need to bring food but should bring our own bedding and water bottle. He set a time and place to pick us up the following day at our hotel.

One stretch early on in the trip out of Agadez was across a dark, rocky plateau. It was unseasonably hot. Our driver pointed out three ostrich up ahead to our left. They were running across the rough jagged rocks like it was smooth pavement. Ostrich were quite common across the Sahel, but now were in decline and struggling. One local later told me that the oil of ostrich is believed to be a remedy against rheumatism. The rumor was that there was high demand for ostrich oil in the Arab countries in North Africa. Later in 2006, Mariama and I would return to the Air Mountain area and learned that the only remaining ostrich were held captive in the town of Iferouane. I was exponentially less excited to see the captive ostrich within a mud-walled lot than the ones running wild.

As we proceeded northward up a broad sandy valley between dark rocky ridges, our driver started talking about the abandoned town of Asode. He was impressed by the size of the Asode cemetery. "A lot of people died there!" We rounded a bend in the road and there were the ruins! The ruins looked similar in architecture to Agadez, with narrow meandering streets and small often innovatively built adobe houses. But rather than the adobe walls of Agadez, these walls were made up of small stacked stones, assumedly held in place in the past with mud. Nearly all the remaining standing house walls were only waist high or so, but from the pile of rocks at the base, you knew they were taller in the past. Near the center of town was a large building, which I was told was the mosque. Later, in 2006, I was allowed to visit the mosque in Agadez, and I saw a very similar layout between the Agadez and Asode mosques. Our driver had no idea who the folks were that had lived there and where they went. With a little digging, I learned that Asode was active in the 11th and 12th centuries with indigenous folk who were ultimately driven out by the 16th century by the newly arriving Touregs coming from the north. Asode was located in almost the center of the Air Mountain range and per our driver was an important stopping place for much of the caravan traffic across the Sahara.

On the trip, our driver would stop in some of the Air Mountain towns to buy fresh vegetables and meat for our food supply. These small communities were often built along the edge of some of the large washes in the Air system. Shallow wells were dug and provided adequate water for irrigation of dates, vegetables, pomegranates, potatoes, and more. So, we ate well on the trip. One such community was Timia, with a high rock and cement structure on a finger ridge between the confluences of two large washes. South of Timia, the wash went over a small cliff. At the base of the cliff was a small pool of clear water. This is called the Grande Guelta. We waded and some swam in the cold water. The air temperatures were not quite warm enough for any prolonged swimming though.

As we returned toward Agadez, we pulled off the highway from Arlit to Agadez on a sandy two-track road. Our driver said we were stopping for lunch. After we unloaded, the driver told us there were large, life size petroglyphs of giraffes up on the rocks around us. He invited us to explore them. With him and the other driver setting up shade and food for lunch, we went rock climbing and exploring! Sure enough, there were two full-sized giraffes carved into the top of several of the larger flat-topped boulders. One giraffe even had a lead rope around its neck.

All too soon, we were back in Agadez. We spent a night in Agadez before heading home. I was sad to leave Agadez and the Air Mountains, but looking forward to getting back to my family and new daughter.

— 23 —

BUFFALO (1991-1995)

My last trip to Niger for work was in 1991. At that time, I estimated I had spent a total of 10 years working in Niger between 1979 and 1991. There were multiple trips back and forth across the Atlantic. Toward the end, it seemed I was being asked to do a year's worth of work in only the three-month rainy season. I kind of dreaded going back to Niger, as I was consistently overworked. I had moved my family in 1988 to Las Cruces, New Mexico, where I was finishing my PhD. The project in Niger still wanted me to return, but in the end the USAID funds to cover my salary were "borrowed" by a creative Niger government accountant who needed to pay Niger civil servant salaries. After two months of promising to send funds, but none arriving from Niger, New Mexico State laid me off. I was able to find work in the Range Department there while I finalized my dissertation. I began to frantically look for work. I finally landed a post-Doc position with the USDA Agricultural Research Service in Fort Collins, Colorado, mapping nitrate leaching in agricultural lands along the Front Range. After three years, the post-Doc position was winding down and I applied for a Research Associate position with Augustana College as a visiting scientist at the U.S. Geological Survey EROS Data Center in Sioux Falls, South Dakota. This new project was using remote sensing to assess bison and fire interactions on Nature Conservancy bison ranches across the Great Plains.

As part of my interview, I joined the Augustana College and Nature Conservancy staff doing pasture sampling on a Nature Conservancy bison ranch in the northern Sandhills of Nebraska, not far from Stuart, where I had graduated from high school. The field data we collected was to support a remote sensing mapping of pasture production that I would be doing if I was a successful candidate. I applied many of the things I learned in Niger on collecting ground truth data, only this time it would be a much higher resolution satellite. The Niger work was at a 1-kilometer pixel size and this bison work would be at a 20-meter pixel size.

Often while sampling grass production out on the hot prairie, herds or solitary bison would wander past. I pretty much treated them the same as I would cattle, but with an added degree of caution. We were more concerned about poison ivy than the bison. We were diligently clipping plots a half meter in area, which were later sorted into live and dead components, oven dried, and weighed. Further analysis quantified leaf area index and chlorophyll content. These clipped plots were measured with a radiometer to get the spectral reflectance of the plot in multiple spectral bands in the visible, near infrared, and shortwave infrared reflectances. From these reflectance, vegetation indices were calculated and correlated with the clipped plot observations of biomass, chlorophyll, and leaf area index. The handheld radiometer also sampled the 20 by 20 m area surrounding the clipped plot to scale up the plot attributes to the resolution of the satellite measurements. Correlations between 20m plot vegetation attributes and satellite spectral vegetation indices would allow regional mapping of the vegetation biomass, chlorophyll, and leaf area index. Enough science mumbo jumbo.

We worked hard, and after starting work with Augustana College, many return trips and trips to other bison ranges in Kansas and Oklahoma were conducted and several journal manuscripts were published. In the northern Sandhills bison ranch, our crew was asked once to help with the bison roundup in the fall. The bison ranch crew rounded up the bison and drove them with horses through a gate near a corner in the large, extensive bison pasture. Some of the bison made swipes at the horses with their horns, and occasionally the herd would just charge back through the horses. This made the horses very skittish and nervous, so in later years the bison crew relied more on four wheelers in the roundup. This made the bison hate four wheelers. The corner gate led into a smaller pasture which funneled the bison into a corral. Using a series of gates on smaller and smaller corral sections, the bison would be forced into the pinch chute. Once in the pinch chute, the bison were vaccinated and some of the young heifers were gated up the loading chute into the truck for sale. The big cattle truck semi backed up to the loading ramp with the diver closing off forward compartments as they filled with bison. Our job was to flush bunches of bison forward toward the chute and loading ramp. The tool we were given to do this with was an empty aluminum beer can on the end of a stick with a rock or something inside the can. We were to rattle the can to scare the buffalo forward. This worked on 95% of the bison, but there was one young female bison who had a crazy look in her eye, much like that crazy cow back in Africa. This bison had broken the horn covering on her right horn, so there was just a bloody boney projection on one side of her head and normal horn on the other side. This bison was oblivious to the rattling beer can. She was the last one, and we could not get her to move forward toward the chute. I thought I could jump up and climb that corral fence pretty fast, and I was getting pretty fed up with this bloody horn bison. So, I jumped down into the middle of the corral with beer can on a stick, blocking the bison from retreating further toward the back of the corral with my body and outstretched arms. The bison stopped and faced me and stared (same thing that crazy cow in Niger did before she charged), unafraid, and with a touch of crazy in her eye. I stood my ground and rattled my beer can confidently. However, at the same time I kept the corral fence in my peripheral vision and quietly and quickly planned my impending and seemingly inevitable demonstration of corral fence rapid ascents. A question that kept going through my mind was, "Who's idea was this?" After what seemed like 15 minutes, but probably was only 45 seconds, "bloody horn" turned back and headed forward toward the chute. Afterward my colleagues also questioned both my logic and sanity on the decision to jump into the corral with that crazy bison. So, assertiveness can pay off, but you should have a plan "B" well thought out before you execute.

— 24 —

THOUSAND-YEAR-OLD EGGS (1996)

My last assignment as a research assistant for Augustana College was to assist on a research trip to northern China and Mongolia to help assess differences between North American grasslands and Asian grasslands. North American grasslands tend to have more warm season, or C_4, grasses than Asia at comparable latitudes. This appears related to the orientation of the mountains blocking southern migration during cold spells like the ice ages. Mountain ranges in China and Mongolia tend to be oriented east to west which would block or constrain southern plant migrations during cold periods while the U.S. mountain ranges, primarily the Rocky Mountains, have a north to south orientation, which allows fairly unrestricted southern retreats of warm season grasses in the Great Plains during extended cold periods. Our goal was to collect grass and soil samples for analysis that would quantify both current and historical warm season grasses in eastern Asia.

At this time, I had accepted a land cover mapping position for a government contractor at the USGS EROS Data Center, but was asked by Augustana College staff to delay transitioning to my new position until after this field trip. I agreed once my future employer approved. The trip was planned for about 2 weeks in the field in northern China, in "Inner Mongolia," followed by another two weeks further north in Mongolia. Arrangements had been made with scientists in both countries to assist in our sample collection, in particular, specialists familiar with the prairie species we would encounter.

We spent about a week in Beijing meeting officials, preparing our sampling equipment, and working out final details of our route. The word was that because of the rains, many bridges were washed out and not passable to the southern Ordos Plateau where our sampling was to begin. Our goal was to cover a fair range in latitude with our Asian prairie sampling. As the departure date arrived, we received word that there was a passable route but it involved a significant detour. We had two four-wheel drive vehicles consisting of one Toyota land cruiser or similar and a Beijing Jeep. I preferred riding in the latter. Somewhere along the way we picked up an elderly Chinese fellow who was the botanist. He struck me as a very traditional man, and he knew all the plants we brought to him, which impressed me.

One river crossing, I remember having to go off road following nothing more than some tracks across farm fields followed by a dicey river crossing. I could tell by the driver's look of angst, the high water, and deep sand that this was a sketchy crossing at best. Both vehicles made it through the crossing with a huge sigh of relief. We pushed on along bumpy roads, arriving at Yulin well after dark. We were hosted by officials from the town of Yulin and were introduced to the tradition of bringing out the dishes of what I would call the toppings and only bring the rice out much later. This was so the guests could fill up on protein and not mostly just rice. I quickly tired of this tradition after the first week of our field sampling. I preferred the mix of rice and toppings rather than munching on just the toppings.

We headed north out of Yulin, toward Baotou and Houhehut on the Yellow River. This is where we sampled our first set of plots, getting plant biomass, soil samples, and a soil description. The soil organic matter would be run through a mass spectrometer to determine the prevalence of warm season grasses historically through carbon isotope analysis. The current vegetation samples would be analyzed in a similar manner and, along with our species composition data, would quantify current warm season grass composition. Along the dusty road, I saw a ring-necked pheasant in its wild native homeland. Back home in South Dakota, we had a lot of these Chinese birds around. I was impressed with how all the grasslands we were seeing on the Ordos Plateau were so heavily grazed. I was amazed a pheasant was able to find enough cover to survive. But this was the early to mid-rainy season, so we were seeing the grasslands at their worst, after a long dry period with little growth.

The distant smokestacks billowing black smoke high into the air told us we were approaching the town of Baotou on the Yellow River. The Yellow River was heavily laden with sediment and had about the same color as the dirt. Our group headed east toward the town of Houhehut. One of the lead American scientists in our group had spent some time in Houhehut before and had some academic friends there he wanted to visit. Also, it was the home town of the Chinese botanist. He invited us all to his house for supper one of the nights we were there. His apartment was small but room was made for a long table. His wife served us a variety of food. One in particular was what looked like hard-boiled eggs but with a blue glow where the yolk should be. The botanist invited the lead American scientist to have a sample. He graciously declined but said that I would happily have one of the eggs. Their name translated to "thousand-year-old eggs." I certainly did not want to insult the elderly botanist, so I helped myself to one of the eggs, all the time cursing the lead scientist in my head for putting me in this awkward situation. I waited until a good conversation was going before sneaking a taste. It was very strong, kind of like blue cheese on steroids, with a strong sulfur aftertaste. Now I love blue cheese, and I felt challenged to eat this egg without showing any displeasure. I kind of staggered bites of the sulfury egg with other, more palatable items on my plate. I finally finished off the egg, without throwing up or making a grotesque face. It was interesting to see the cramped living conditions (but the apartment appeared fairly modern) of fellow academics in China.

We departed Houhehut and headed east, northeast to the town of Xilinhot. We crossed the low mountains north of Houhehut, which had a few scattered clumps of coniferous trees. Chinese scientists on the team said historically these mountains were covered in trees but harvesting of wood had depleted the forests. Current efforts were focused on some re-forestation. After passing through the mountains, we proceeded out across a grassy plain. There was more grass here than on the Ordos Plateau. The moderate grass cover and heights were similar to the mixed grass prairies of the U.S. Great Plains. After a long day in the hot vehicles, we arrived late in the afternoon at Xilinhot. I was thirsty and hungry. Protocol was that we had to go meet the mayor and some of his staff. After some pleasantries in the mayor's office, he offered to join us for supper and recommended a restaurant. Protocols continued as the mayor launched into a series of toasts with freely poured Chinese wine, which is really strong. After about three or four shots, I decided I had enough. I whispered into the lead American scientist's ear that I could not do many more shots. He said to just sip the wine, no need to go bottoms up on a toast. This worked fine as I was able to survive until the food arrived. The dishes full of sauces, meats, and other toppings were loaded on to a big "lazy Susan" in the middle of the table, which could be rotated by anyone at the table to access his or her desired food. The protocol of not bringing the rice out until much later was carried through yet again.

The next morning, we parted for a research station where a lot of grassland research was going on. We arrived in a productive and very lightly, if at all, grazed system. Out in this grassland was a set of buildings which looked like several dormitories, offices, labs, and garages. The lead U.S. scientist and I shared a room with two single beds. The next morning, we set out to visit a pristine prairie site. When we arrived, there were several ladies out collecting flowers. It turned out that this area was one of the most diverse prairies on earth. Of particular notoriety was the high variety of flower species that were everywhere. This was the kind of pristine, productive prairie that was needed for our study. Our crew continued our sampling protocol of soil pits, vegetation samples, biomass sampling, and vegetation composition. We returned to the research station for supper and then the college students that were there brought out a boom box and held a dance out on the parking lot. I learned swing dancing in college and still remembered most of my moves. There was a good song and I asked one of girls to dance. Everyone just stopped and watched us dance. Afterward, I was out of breath, and the next song was not to my liking, so I sat down next to one girl on our team who did a lot of translation for me. There was a line of college girls waiting in line to dance with me. I danced with a few, but most of the songs were not to my liking. I also played some ping-pong with one of the male students and thought I did OK. But the next game he brought his "A" game, and I think I was only able to score a couple of points in the game.

We now headed south back to Beijing, but we passed through an area were sand dunes were advancing on farm ground. I remember one abandoned house which was half buried in a sand dune. We proceeded back to Beijing to prepare our samples in the lab and prepare for another two-week trip up in the Mongolian grasslands. We flew out of Beijing for Ulaanbaatar, the capital of Mongolia. We were met by some of the team of scientists who were going to be working with us and checked into a modern hotel. We spent the next couple of days getting ready to go out for two weeks in the bush and meeting the various scientists that would be traveling with us. Ulaanbaatar had a classic Soviet mass production architecture, which resembled dormitories made out of slab concrete. We visited a tourist-oriented restaurant one evening. After a very good meal, we were entertained by a traditional Mongolian singer. It was a low guttural sound that had at least two harmonic tones. It sounded surreal and nonhuman and fascinated me.

We had two Soviet-built four-wheel drive vans called Yaks. These things were serious four-wheel drive vehicles. All the roads were gravel or two-track once we were about five miles out of Ulaanbaatar. The first major pass we summited, we stopped. There was a large pile of loose stones with some empty whisky bottles strewn about on top. There were some colorful little flags similar to those I have seen in pictures of base camp at Mount Everest and in Tibet. The drivers started walking around the pile of stones slowly, picking up small stones on the ground and throwing them on the pile. We were told that this was to bless our trip and to help ensure good luck. The entire team proceeded to follow suit, picking up rocks and tossing them onto the pile of stones.

I was struck by the similarity of the grasslands here in Mongolia with those I knew in the western U.S. and West Africa. I was also impressed by similarities with the pastoralist people I have seen and known around the world. We saw some herds of animals, and we saw a lot of the pastoralists' empty winter quarters, built in a crescent shape around a manure-filled center. Reportedly, some of the pastoralists return to their winter quarters in mid-summer to cut and store some hay for winter. Typically Mongolian winter snows are either not as deep or as persistent as they can be in the northern Great Plains where hay is needed to carry livestock through the winter. Because of this winter hay was kind of seen as optional or a luxury in Mongolia. Interestingly, a year or two after our trip, Mongolia had a severe winter with deep persistent snow which decimated the livestock herds and significantly impacted the pastoralists' livelihoods.

We saw occasional camels, Bactrian two-humped camels. These were larger than the camels I was familiar with in Africa. The grasslands looked in good condition and lightly grazed. This was probably because we were traversing what appeared to be their wintering area. We came across another scientist group doing research. Apparently, we were looking for their camp and were expected. They let us stay in one of their canvas wall tents. The only problem was that the tent had no floor and the mosquitoes freely accessed our sleeping quarters along the bottom edge of tent. Fortunately, it was cold at night and that helped keep the mosquitoes at bay.

I think it was about the 6th day out of Ulaanbaatar that we finally came across a fairly substantial town. At Baruun Urt, we bought supplies and some souvenirs in the market. One of the U.S. scientists was wearing a little day pack around the market. The day pack was full of money we needed to finance our trip (gas, food, lodging, etc.). This scientist had a swarm of local people hovering around him, like what often happened to me in an African market. Somehow, the locals did not hover around me and I was perfectly happy about that. I took pains to position myself where I could keep an eye on that day pack, and the scientist casually shopped around for souvenirs and joked with the folks hovering around him. He told me later that the way not to get robbed is not to be nervous about the money you are carrying. I might have been a bit suspicious, too alert and focused on the day pack, possibly even perceived by the locals as his (wimpy?) bodyguard.

Leaving Baruun Urt, we headed a bit further east and then looped down into the Gobi Desert, paralleling the Mongolia-China boarder. As the climate got drier, we noticed dark-colored hawks everywhere. I think there was a healthy population of grasshoppers or locusts that the birds were feeding on. It was the highest density of hawks I have ever seen. The birds were perched on the ground, 100 to 200 yards apart. We proceeded on in the heat until we came to the southern rail town of Erenhot (also known as Dzamin Uud?). We needed to restock with both gas and water. The town seemed half vacant in the hot afternoon sun. The electricity was off, as the town was trying to conserve diesel fuel used by the generators. We hung out at the gas station for what must have been the siesta period. Finally, the power came on and we filled our fuel tanks and headed north back toward Ulaanbaatar. The vegetation became denser and greener as we advanced northward.

I think it was in this stretch where we started looking for another scientist group we were supposed to rendezvous with. We finally came upon a huge camp in the prairie. There were six or seven vehicles and a circle of little dome tents surrounded by larger lab type tents. This was a Japanese research team lead by Sashi Honda (sp?). He had soil scientists and a remotely controlled helicopter with live video and multispectral imaging. There must have been 30 or so of students and professors. About half of the students were from Mongolia and half from Japan. They invited us to spend the night. We ate and drank beer (or was it wine?). Anyway, I was feeling pretty good. Our little team was sitting together when Sashi announced that we were going to have a singing contest. I hoped that we would be excluded from the competition, but Sashi divided us into groups based on nationality. There was a single Chinese woman, who declared that she was perfectly comfortable competing solo. It turned out the Japanese had sheet music. I asked to borrow one of their music sheets, hoping we could at least get the lyrics right. Unfortunately, it was all in Japanese. Let me tell you the American team was sweating blood. None of us were musically inclined, and all we could come up with was songs like "Old McDonald's" and "Home on the Range." The Mongolians sang loudly, apparently some traditional songs. But the show stopper was the Chinese lady who broke out singing a Chinese opera song or two or three. She won. I was feeling so good from the booze and concentrating on trying to sing and remember lyrics that I essentially stepped in my bowl of food.

The next morning at breakfast there was a Japanese student who wanted to know if one of us left a fecal deposit in the big hole over the hill. We all looked confused and stated that we had not. He was the soil guy and someone had pooped in his meticulously dug soil pit. He took us over to see his soil pit. It definitely was a human feces. The soil pit had nearly perfect vertical walls with 90 degree corners, a master piece! No one ever owned up to the deposit in the center of the pit, as we saw the soil guy carry on questioning other students.

Mid-morning, a pastoralist came by and was letting some of the students ride his horse. Or should I say pony? All the students seemed to be extremely novice horsemen. I tried to get a good look at the horse and his tack. I am 5'11" and felt like a giant, but the pastoralist asked if I wanted a ride. I climbed on board, and the stirrups were so short that I felt like a race horse jockey. This was simply untenable. I stood up in the stirrups and sat behind the saddle cantle, where a passenger would normally ride. Folks chuckled, but now I felt comfortable and took off at a gallop over the hill and back. Many years later, I met Sashi Honda in Kansas with this crew of students and the remotely piloted helicopter doing grassland research. A couple of years after that, I heard that while doing research in Mongolia, he sent some of his students off to town to get some takeout food. On the return trip, they had an accident and one or several students were killed. This impacted Sashi intensely, as he personally apologized to the dead students' parents back in Japan. Sashi apparently was so impacted by the students' deaths that he never did any more field work.

We departed the Japanese, and the road climbed through several mountain passes as we approached Ulaanbaatar. Once in town, we had to prepare our samples for shipping to Beijing. I think we spent two or three days in Ulaanbaatar before flying back to Beijing. Once in Beijing, we further prepared our samples and paperwork needed to bring our samples back into the States. There was some concern about some of the vegetation samples. As I recall, we had approval to bring back soil samples, but not vegetation samples. While in Beijing, we were invited out for supper and Beijing Duck was served. This is a delicacy and was very tasty. I was getting somewhat competent with the chopsticks by this time. As the plate of duck went by I used my chop sticks to grab what I thought was a drumstick. Boy, it must be a bit overcooked, I thought, as I could feel with the chopsticks that it was very hard. I started chewing on it and I could not figure out what it was. It was really bony. The Chinese female scientist next to me said, "The head is hard to eat, isn't it?" I was shocked but did my best to dead pan and just nodded and kept up my attack on my "drumstick."

Our flight home was uneventful, but as we went through U.S. customs, we were a little anxious about our vegetation samples being seized. As I recall, the lead U.S. scientist had me put some (most?) of the vegetation samples in my suitcase. The agriculture inspection guy walked by with his beagle with the green jacket. The dog paused briefly beside me. I smiled and started to reach down to pet the dog with a big smile on my face. The officer said, "Do not pet the dog." Then the dog walked on with officer following.

—— 25 ——

CENTRAL ASIA (2003)

The lead U.S. scientist on the China and Mongolia trip had begun working at the USGS EROS Data Center in Sioux Falls where I worked. He wanted me to work on a project he was aware of in Central Asia. He had asked me multiple times and I considered it, weighing the benefits and disadvantages. Finally, I agreed to work on that project with him and reduce my time on a national land cover mapping effort (NLCD). I think it was the intrigue of doing remote field work in foreign grasslands that tipped my decision. In working with the land cover mapping effort, I had become interested in something called Decision Trees and Regression Trees being applied to remote sensing integration with digital map products (Geographic Information Systems or GIS) and was having good luck with these "machine learning" techniques. The Central Asia project task was to extend carbon flux measurements from carbon flux towers in Kazakhstan, Uzbekistan, and Turkmenistan spatially and across multiple years using remote sensing (see http://dx.doi.org/10.1007/s00267-003-9156-8 for details). I had the tool (Regression Tree) that fit this problem of carbon flux mapping, and I had the remote foreign grassland field experience.

Several preliminary steps needed to happen to allow the Central Asia flux mapping to proceed. One was verifying that satellite observations could be associated with carbon flux measurements. I did this using a flux tower in DuBois, Wyoming. A second task was to select and prepare the satellite data I would be using, along with any other digital map products that may help in estimating carbon fluxes. We were interested in the collapse of the Soviet collective farms that occurred with the end of the Soviet Union. Wheat farming in Kazakhstan in particular had been largely abandoned when Soviet subsidies dried up. Collective farm equipment and livestock were used to partially pay off collective farm workers for past wages and whatnot. As a result, most of the livestock got eaten, so the rangelands were in great condition, being only lightly grazed. Using grass carbon flux mapping, we sought to quantify the land cover change carbon impacts associated with wheat farming converting back to grassland.

The project funded a visiting scientist, Sasha, from Kazakhstan to come work with me in Sioux Falls. After working together getting the satellite and GIS data together, a field trip was planned to Kazakhstan. I had proposed the idea of a second flux tower in Kazakhstan, which could be moved to a different land cover or a different location every couple of weeks or every month. This assumed that the satellite data would allow us to make flux estimates in the time periods when the flux tower was absent at a site, thus allowing one flux tower to monitor two or maybe three sites.

A fellow remote sensor from EROS, Brad, would be joining me. Brad and I flew to London and then on to Moscow. The field crew in Kazakhstan wanted us to bring a battery-powered electric drill that would facilitate moving the roving flux tower. My bags (all carry-on bags) were full of clothes and equipment I needed for my work. When I got this last-minute request, I reduced some of my personal stuff and repacked more efficiently such that I could take the drill as a carry-on. Checked bags were a liability because our flight to Moscow from London was on Aeroflot, which had no baggage agreement with our flight to London. A checked bag would mean we would have to go out past security to the baggage claim, get the bag, recheck the bag at the Aeroflot ticket counter, and then go back through security. We had heard that the lines were long at the London airport, so we expected our time between flights was going to be just too tight for checked bags. However, the plan fell flat on its face when airport security in South Dakota would not allow the drill to be taken as a carry-on. I was forced to check the bag at the last minute. I was cursing that this drill may completely foul up my flights and I may arrive later than my traveling companion, Brad.

On the flight to London, I tried to sleep as much as I could. Brad and I walked quickly through the London airport, headed to baggage claim. There was an escalator going down one story with a long line of people waiting to get on. I led Brad, skipping rapidly down the stairs, passing 30 to 50 people in line and on the escalator. Luckily the drill was quickly available at baggage claim and there was no line at the Aeroflot check-in desk. The lack of a line at Aeroflot was both a blessing and a concern, as we knew Aeroflot had a poor safety reputation. Somehow, we got through security and to our Aeroflot flight gate in time! I was immensely relieved but still a bit nervous about the Aeroflot flight. The flight was less than half full, and the interior looked OK. When we took off, however, several of the ceiling panels popped out and sagged down. This was not alarming, just disconcerting. The flight attendants walked down the aisle and calmly and mundanely lifted each ceiling panel easily into place as if they had done it 500 times before. The flight to Moscow was uneventful, but I had my nose to the window looking at the towns and the countryside. The towns tended to have what appeared to be miniature villages on their outskirts. I later learned that these where small private garden plots, each garden plot often with an adjoining small cabin or shed. These small gardens were started with the Private Garden Plot Act and produced a substantial part of Russia's food production in the late 1900s. We landed in Moscow with the now usual ceiling panel descents.

When we stepped off the plane and into the terminal, we were in a different world. The signs were all in Cyrillic script. I had thought reading French signs was a pain, but now I was nearly completely helpless. Every word over the intercom was in Russian as well, so Brad and I were kind of flying blind. We did manage to find the baggage claim and a taxi to our hotel. Brad and I toured downtown Moscow the next day, wandering around down by the Kremlin and other places. It must have been the next morning when we caught our flight to Astana, Kazakhstan, and we were met by Knot, our collaborating Kazakh scientist, and driven to Shortandy, which was an hour or two away. He dropped us off at a small bed-and-breakfast place. We had separate, clean rooms with a single bed, but shared a common bathroom. Adam, a graduate student from California, was also staying there. Some other local visitors would sporadically occupy the other two remaining rooms and would trash our shared bathroom. It was unclear if this was to insult the Americans or if this was natural in this culture. Previously, we had some visitors from Russia and Central Asia come to EROS one year. There were emails circulating at that time about the men's restrooms having chronic cleaning issues throughout the conference. But the same question persists, custom or international tension? Maybe a bit of both.

Adam, co-resident with us at the B&B, was a graduate student working on the project primarily insuring the flux towers were operating correctly and getting their data downloaded and processed. Adam spoke what appeared to me to be fair Russian. We visited the site of the stationary flux tower, whose data we would primarily be using in our mapping. Then we proceeded to witness the moving of the roving flux tower. They had a crew of about six that included Adam, and it took them about an hour to get everything disconnected and ready for transport. Adam seemed in control and giving advice or orders to the others to facilitate not damaging the short tower or its equipment. They had the tower staked down with hefty tent stakes so it would not get blown over by the wind. One younger fellow on the crew was trying to get the stakes out of the ground by digging the dirt away near the top of the stake and wiggling the stake to loosen it. I have put up and taken down my fair share of tents, so I have developed an efficient approach for pulling stakes. I walked over and grabbed a wrench lying on the ground. It was a typical open end and box end wrench combination. I walked over to the stake the fellow was trying to dig out and slowly reached down and put the box end of the wrench over the end of the stake with the box end angling downward. Then I slowly pulled up on the handle of the wrench and up came the tent stake. The fellow grabbed the wrench from me smiling and nodding and went off to remove the other remaining stakes. Knot, the lead scientist said something to the fellow and the fellow replied. Then someone translated for me that Knot had complimented the fellow on his ingenuity which resulted in a useful solution. The fellow had replied that it was me who had showed him. Poor fellow. I wish he had just taken full credit, as I certainly did not need it.

We started getting our equipment ready for our land cover ground truthing field trip while Knot found a rental vehicle and driver to take us out on some of the roads we hoped to drive. We had GPS, which was linked to our laptop. The laptop had satellite imagery on it showing an arrow where the current position of the GPS was on the image. Our image had pixels that were 30 m (about 30 yards) on a side, so we could see many of the roads on the zoomed in image. We wanted to test our setup in a car, so Knot got his personal car and kind of drove us around Shortandy and vicinity. One minor complication: to avoid suspicion that Brad or I might be spies, Sasha had to run both the laptop and the GPS. I could help him troubleshoot some of the problems we had only verbally. I was not to touch either the computer or the GPS. I caught myself several times reaching for the GPS or the laptop when I was actively involved in resolving an issue, but stopped myself. Finally, Sasha had it up and running while we were riding around in Knot's car. We headed back to Knot's lab. Sasha was up front in the passenger seat and we were talking excitedly on how accurate the GPS and imagery were. I said something like, "Why you almost do not need to look out the window. One could almost drive the car by looking at the laptop screen." Anyone who knows me knows they should expect satire at any moment, but I looked up and Knot was about to have a head-on collision with an oncoming car on a Shortandy street! I yelled, "Watch out!" Knot looked up from the computer screen and quickly adjusted his steering. I pointed out that while we could see the road on the satellite imagery, we could not see the traffic in real time. (Again, my attempt at humor could have, in hindsight, been interpreted as derogatory.)

With our equipment in place, Sasha now could log the actual land cover on either side of the road as we drove along. Sasha and I had previously picked some routes that we thought would be informative. The next day, Knot introduced us to our driver and assured us that this was the best vehicle in Shortandy and that he knew the route we wanted to take. Great! Sasha, Brad, and I jumped in and we were off.

We saw lots of abandoned collective farm headquarters and fields. The grass looked to be in excellent condition. Throughout our travels in rural Kazakhstan, we saw very few livestock. We were proceeding well along our route and about two-thirds of the way to the first major village and major road junction. There were mud puddles in the road and the car we had was more of a small highway car, but the driver was unconcerned and seemed to be doing just fine navigating the obstacles and keeping us from getting stuck.

Brad had been complaining about the smell of gas for quite a while. I concurred that I smelled it too. Our driver was chain smoking and had his windows down so that might have diluted the gas odor. Brad mentioned the smell several times more before he emphatically stated that his butt was wet! We had the driver stop. Brad's pants were dark and moist with a strong odor of gasoline. The rear seat cushion on Brad's side was wet with gas. I was scratching my head and asked the driver if I could see in the trunk. I could see no source of the gas. Then I asked if I could remove the rear seat cushion, which revealed a shallow pool of gas under the cushion on Brad's side of the rear seat. There was a removable drain plug there that I showed the driver and asked him if I could remove it, so the gas could drain out. He said yes. We decided that we would put the seat cushion back (once it dried in the sun a bit), leaving the drain plug out (dust was better than gas). Then we removed the rear floor mats and put them over the seat cushion. I volunteered to sit on that side as Brad already had been exposed to this precarious situation. Brad said it was OK, he would stay sitting on that side. Obviously, also no more smoking by the driver. We piled back in and in less than a quarter of a mile, were hopelessly stuck in the mud. (When it rains, it pours?)

After exhausting ourselves pushing and with no shovel to dig (there was absolutely nothing in that trunk other than a jack and spare tire), we used maps and the satellite image linked to the GPS to estimate the distance to the village ahead. I think it was 2 or 3 miles, nothing insurmountable, and it was a cool, sunny day. Sasha, Brad (with his dirty butt pants), and I started walking. I (or Brad?) had some cash which Sasha and the driver thought should cover the costs to get the car pulled out. It was a long walk, and I am pretty sure we critiqued our driver and his car pretty thoroughly as we walked along the abandoned wheat field two-track, rutted, and muddy road. As we were starting to tire, we arrived in a town that appeared to be 90% vacant. Maybe everyone was hiding from the guy with the dirty butt who smelled like gas? Sasha finally saw a couple of young men, talked with them, and then headed off to talk to a guy who had a tractor in town. In half an hour or so Sasha returned riding on a tractor along with the tractor operator. Sasha explained the price and we agreed. Then Sasha was off with the tractor operator. Brad and I agreed to just wait in the ghost town until our car arrived.

As Brad and I sat there in a town with probably no one who spoke English, I kind of felt a bit vulnerable. But nothing we could do about it, so we sat on a dirt street leaning against a building wall. We saw no signs for a restaurant or anything. For the 1 1/2 to 2 hours that we waited, I do not think we saw a vehicle or a single person. Maybe the word was out that there were crazy Americans in town?

When Sasha, our car, and our driver returned, they said that after the car was pulled out, they were able to drive it all the way to town without getting stuck again. We looked at the clock and we had covered only one side of the isosceles triangle road network that we had planned for the day. Given that we did not want to backtrack over the muddy road that we just came down, and that reports from the tractor driver were that the remaining roads on our route were major roads in good condition, we opted to proceed forward at full speed, forgoing the land cover GPS data collection so we could get back to Shortandy just after dark. At Shortandy, Brad and I were dropped off at the B&B while Sasha and the driver went off to inform Knot. Obviously, we would need a different car (and hopefully driver) for our next excursion, which was going to take four days.

We shared our stories of the day, and Adam said he wanted to come with us part of the way on our upcoming trip. He wanted us to drop him off at some town on the way and he would catch the train from there. I wondered how he could possibly want to ride with us, given our recent run of bad luck. Maybe a break from the daily grind is better than bad luck?

The next day Knot let us know he had a good car and driver to take us to a distant town. From there we would have to secure our own vehicle somehow to proceed on our sampling trip. This seemed rather undefined, but both Sasha and Knot were both happy with it. I think our departure was maybe mid-morning, and our new driver was expecting to have us to our drop off point before dark. This was a travel day, as our sampling was to be out of our new destination (I do not remember the town's name.) Things started well. The road was good, even paved. After a couple of hours, we got to the town where Adam was going to get on the train. We stopped, he bought his ticket, and said that train would be along in 15 minutes or so and we could leave. I suggested to Brad and Sasha that we could spare 15 minutes, as I was kind of curious what the train was going to look like. Having lived in third world countries and seen their mass transportation, I was still trying to calibrate where this place fell between a developed country and a third world country. Both Brad and Sasha agreed to wait, knowing our road ahead was a long one.

It seemed like we waited an hour for the train, but probably was only 30 to 45 minutes, but definitely not 15 minutes. Adam had his stuff as the train stopped, but he just stared at the train. People were standing in the open doors to the cars and hanging out the open windows. The impression I got was that it was filled to the gills or it was just hotter than Hades in there. Adam decided he was not going to get on the train, but travel with us instead. Adam went back to the ticket counter and the guy that sold him his ticket refused to give him his money back. They argued, waited, and argued some more before Adam was refunded a portion of his ticket price. Adam threw his stuff back in our car and we were off.

Not too far down the road (30 minutes?) was a detour sign pointing out across the grassland beside the torn-up road. There was no effort to blaze a path through the prairie at all, other than to hang that detour sign up and then they were done. It was bumpy and very rough in places, with the bottom of our car dragging and banging many times. Oh, and the detour? It lasted for the next 4 or 5 hours! Dusty, hot, and bumpy! There was some grumbling from us, and Sasha said the driver had no idea that this road was under construction. I think he talked to someone in a vehicle going the other way and learned that it would be like this all the way to our destination town. He estimated our arrival at midnight to 1 a.m. I bet Adam was now wishing he had gotten on that train!

We finally pulled into our destination town after midnight. The only folks who appeared to be up were the police at the police station. Sasha was telling them our objective, our need for housing, and our need for a rental vehicle. One of the policemen said his father had a car and could possibly put us up. We followed the policeman on his motorbike and waited while he went in to talk to his Dad in the middle of the night. By this time, I was starting to lose confidence in relying on other people's confidence to make decisions, but, hey, that ship had already sailed.

Miraculously, we were taken in by the policeman's father. We said good-bye to our driver and he headed back to Shortandy. Our host had a fairly large and fancy house. I suspected it was adobe covered with a thin coat of cement. We all had to sleep in one common living or entertainment room with a couple of couches and two of us sleeping on the floor, but beggars cannot complain. We had not eaten supper, but no one said a word about that.

Morning came and there were some wisecracks passed among us as we woke and got dressed. It turned out Adam was both witty and had a good sense of humor! Our host was a tall, tan-skinned man. He had an easy and friendly smile and was probably in his late 40s. His wife was an attractive woman in her early 40s with lighter skin. Our host said she had lived in the capital city of Astana, much further to our south. He bragged that she knew English, and I asked her a few questions while we were there. She initially tried to answer, but soon just stated that she no longer knew English.

Breakfast was hot tea and some crackers or graham cracker like things along with some bread and jam. Adam warned me when I tried a cube of sugar in my tea that sugar was addictive, it was a slippery slope.

Always curious, I noticed that the fuel his wife used for cooking and heating was dried cow manure. She seemed irritated that I noticed. The toilet was an "outhouse" away from the house and near the gate to the road. The "outhouse" had gaps between the boards that were about 4 inches apart. This way you could see and greet your neighbor coming down the street while you were taking a dump.

Inside the walled yard was a fair amount of old, dried cattle dung (their source of heating fuel?). I think our host said the cows were out on their summer pasture, too distant from town to return home nightly. Folks from the town paid a guy to take care of their livestock. Apparently, they save the grass near town for livestock feed through the winter.

Our host mentioned that he did have a car and could drive us. Sasha and Adam were translating for us. Sasha described the route we were interested in. Our host said that was a good road and he could take us there. We also asked about a return trip to Shortandy, knowing the road was terrible. Our guest said there was another good, paved road that would get us to Shortandy and, yes, he could drive us there as well. We were all very happy with this news and his price was reasonable! Maybe Adam had memories of our trip out here about how things sometimes turn out to be not as promised or not as we envisioned, and he suggested we actually look at our host's car. Instead of being upset about this implied question of authenticity, our host was excited to show his car! We walked a block or two and then he unlocked some double doors with two sets of locks. And there was a mint condition old Volga car. We were completely satisfied with his vehicle and set off the next day sampling. We returned to his house for another night and then he returned us to Shortandy.

Knott drove us to Astana the evening before our flight departure, and we stayed in a modern hotel there. The flight to Moscow was uneventful, but at the Moscow terminal it was very difficult for us to find the gate for our flight to Frankfurt, Germany. We could see a sign with our flight number on it with some text and a LONG line. We patiently stood in line, with a substantial number of people also built up behind us. It seemed like we waited 45 minutes or so. I remember being concerned that we might miss our flight waiting in this line. Finally, we got to the ticket agent, a young attractive woman, who started firing questions at us in Russian. We must have had that deer in the headlight look as we stared at each other. Before we could say we only speak English (and Housa or some French), she asked us in English if we spoke Russian. "Not a word," was our reply. She asked us to step aside and wait. We stood there as she processed everyone in that line that was behind us. My concerns grew as the remaining line grew shorter. Would there be any seats left? Was she going to send us off to jail until we learned Russian?

She finally turned to us, took our boarding passes and passports, worked on her computer, and gave us our boarding passes. I think Brad kind of figured things out as we were boarding, noting that our seat numbers were very low. We were hoping for first class, but having seen so many twists and turns on this trip, we were not going to believe it until the fat lady sang . . . and then she SANG!! First class all the way to Frankfurt, Germany!! That was the best ticket agent in the WORLD! They brought me some fancy food with some real stinky cheese and caviar, the latter two were pretty much wasted on me since I was a country bumpkin and woefully unimpressed, but the drinks were FREE and the leg room was GREAT!!

— 26 —

GREAT BASIN (2011?)

There is a Central Asian grass that was introduced to the U.S. back in the days of sailboats and covered wagons. They used this Asian grass as packing while shipping dishes across the ocean from China. Then during the Mormon expansion into Utah, it was also used as packing for covered wagon loads. As homesteaders and religious and other westward migrations occurred, many gave up or discarded heavy items along the trail. The Asian grass was dry and light, but it also carried seed. Thus, the wagon train routes may have been an effective introduction point for this newly arrived grass. This exotic grass had a super short growing season, so it could grow to full seed maturity early in the spring, set seed, and die before the native grasses barely got started. This allowed the exotic grass to essentially steal critical early season soil moisture stocks that the native grasses relied upon later in the spring. Another "feature" of this grass is that it provides a fine, dry fuel which promotes expansion and frequency of wildland fires. This pesky grass, also known as cheatgrass or *Bromus tectorum*, has invaded and altered much of the Great Basin's fragile and sensitive arid ecosystem. There are other invasive grasses and forbs in the Great Basin, but none have been as pervasive as this invasive.

In the mid-2000s, the Bureau of Land Management (BLM) observed a die-off of cheatgrass in some areas. One suspected culprit is a fungus called the Black Finger of Death, but the jury is still out as there could be other confounding factors. I was asked to propose an approach for the mapping of cheatgrass die-off, and this proposal was funded. Without going into too much detail, in order to map cheatgrass die-offs, we first needed to map cheatgrass productivity, or cheatgrass percent cover (see http://dx.doi.org/10.1016/j.rama.2016.03.002). To do this, we needed field observations of cheatgrass cover. So, I headed out to Winnemucca, Nevada, to meet up with some fellow scientists, Collin and Spencer, to: 1) see and observe cheatgrass die-off areas, and 2) collect field observations of cheatgrass cover within cheatgrass die-off areas of both dead dieoffs and adjacent healthy cheatgrass. I had rented an SUV after carefully inspecting the undercarriage of all the SUVs available. I was interested in the one with the highest ground clearance. I met up with my colleagues from both BLM and the U.S. Geological Survey in Winnemucca, Nevada.

The first half day, we saw dramatic examples of cheatgrass die-offs. I mean lots of cheatgrass. Then, literally on the other side of a line, none. Our BLM hosts directed us to other die-off sites, which we visited ourselves in the afternoon. The next day, the three of us were going sampling in an area where Collin had a high-resolution satellite image ordered. We followed Collin's sampling protocols, trying to find GPS locations that were selected based on the high-resolution imagery. The sampling was not too difficult, essentially estimating percent cover of various rangeland components (percent bare ground, percent litter cover, percent shrub cover, percent sagebrush cover, etc.). The tricky part was getting dialed into consistent estimates. Spencer had lots of experience doing this. There were some beautiful views with sun on some of the desert hills on a partly cloudy, partly sunny day.

One problem was that roads that look substantial on the imagery were not always passable. Sometimes, access was limited by "no trespassing" signs or locked gates, but mostly it was just gulleying from recent overland water flow. I was driving down a two-track road with a raised, slightly vegetated center, trying to make good time while being safe to both us and the vehicle. Around a bend and down a slope, there was a moderate rock on the edge of the road's center. I estimated that in my full-sized Ford Bronco that I used to own, it would have cleared that rock no problem. I knew this rental SUV was a tad lower but was fairly confident that it, too, would clear the rock. It did not! There was a loud crunch and we stopped suddenly as I got on the brakes quick and the rock also slowed us down. We jumped out and fortunately the rock had impacted on the vehicle's frame, not the oil pan, transfer case, or something else important. I felt foolish, as I contemplated how far we were from a major road. Cell phone coverage was spotty, as many of the mountains had cell towers on them, so one might not have had to walk too far before getting a cell signal. Still, I should not have gambled with that rock and this vehicle's ground clearance.

I enjoyed the remote and arid landscapes of the Great Basin a lot and hope to return soon.

— 27 —

LOST GPS (2009)

This was my first remote, off-road field sampling in Alaska. I was kind of treated like a newbie or a greenhorn by most of the group. That was fine by me, as I fully acknowledged that I had little experience in this boreal forest ecosystem we would be sampling in Yukon Flats, Alaska, up near the Yukon River. We had a fairly large crew of six. Some of the sampling gear was a bit large and cumbersome. They had an insulated tank of liquid nitrogen for flash freezing some of the soil samples which would be processed later for soil microbe DNA analysis.

There were three of us in my "walkabout" sampling group. Dana, a graduate student from the University of Alaska, Fairbanks; Jack, a research scientist and former University of Alaska student; and myself. Our team was flying in first on a float plane. Dana and Jack had back country Alaska experience, and they went in in the first flight with a fair amount of gear. You always flew in with your gear and some food, as you never knew when the plane would be back (mechanical issues, weather, or other).

I was the only passenger in the second load, but we brought more of the supplies and gear we and the larger group would need. We were flying into Boot Lake (it was shaped a bit like Italy). I had poured over satellite and aerial photos of the area around this lake for 4–6 months, pre-planning for this field work. As Boot Lake came into view flying in the float plane, everything fit exactly what I saw on the imagery. The bush pilot, Jim Webster, had me looking out the window of his brown and white Cessna float plane searching for Dana and Jack. Jim was a wiry, late-50s, no-nonsense, "just the facts, ma'am" kind of a guy. He pointed out what section of shore he had dropped Dana and Jack off on, and it was about the second pass that Jim spotted Dana and Jack holding a big blue tarp up for us to see. I was astounded that I had missed spotting them first. They were so obvious, once you saw them down there in the open spruce forest. They were up on a small ridge, a fair distance from the lakeshore. It seemed a long, steep path to have to haul all the gear, but I was the newbie so I kept my thoughts to myself. Jim landed smoothly on the water, as the wind was very light with no waves on the water. As we taxied up to the shore, Dana scurried down to help secure the plane and unload the gear. Once we had the gear out, we pushed Jim away from shore, and in no time he was airborne and headed back to Fairbanks. One more load of personnel and equipment was to be delivered to Boot Lake later that day.

Dana had been working with me via email and phone over the previous three or so months, working out our sampling protocol. She had a lot of field sampling experience and was highly recommended by her boss for providing innovative and effective modifications to the field sampling on the fly. She designed our sampling scheme to be similar to another sampling group she had worked with. I informed her that our sampling would be representative of a larger area on the ground, roughly 90 by 90 meters in size, to be sure it could be linked to satellite data. Dana and I compromised on our final sampling scheme. I finally met her when we were getting the gear ready for the trip. She was very short, energetic, and smart.

Dana and I started hauling the gear up to the camp, and soon Jack joined us. Jack was an athletic lumberjack-looking guy who must have been six feet or so tall and I guessed to be in his late 20s. Dana and Jack were excited about the camp location because it had thick moss (soft padding) and labor tea (an aromatic low shrub). The spruce trees were 7 to 14 feet apart and no taller than 3–10feet tall. There was one issue, though, that had my attention—it was not far (half a mile) from an active fire line. The western third of the lake had burned, and the fire continued smoldering in the moss and occasionally short flames were visible. Dana and Jack assured me that the fire would advance slowly, but I knew from my firefighting days that it all could change with a good stout wind.

We set our tents up and strung a tarp over our kitchen area. There was still plenty of daylight, and Dana suggested we go to our first site. I had picked sampling locations on the imagery and had the coordinates of lots of potential sample locations in my GPS. I whipped out my GPS and there were sites all around us. We grabbed our sampling gear and headed off bushwhacking toward our first site. We had to walk across a small, burned patch on our route. The site was dense, 8-foot tall birch trees. It was thick, dense, tall brush, but our sampling worked well. We were sampling for multiple attributes. We were using a permafrost probe to sample thaw depth (how far down to the top of the permafrost), vegetation species composition (trees, shrubs, ground cover), and basal and breast height diameters of trees and shrubs so we could calculate biomass. Dana did some small clip plots to get an estimate of the ground cover (moss, grass, sedges, lichens, low shrubs) biomass as well. I was getting soil cores that would be used for soil carbon content, soil moisture, and organic layer thickness. I also left my GPS averaging its position while we were sampling to get a more precise location fix.

We returned to camp happy with our sampling scheme and the time it had taken us to complete our first site. Shortly after we returned to camp, the three other team members flew in and Dana went down to the lake to help secure the plane and guide the three new arrivals toward camp. They did not have much gear, other than their personal packs, as most of that came out with me. Mark was the leader of the group and was a guy who liked to have fun. Kim and Kristine where scientists with soil and carbon expertise. The thing that kind of threw me off was that Mark had a leopard skin colored cowboy hat and the gals had goofy hats as well. It reminded me of a dress-up party. After the first day, I think only Mark retained his hat. Mark, Kim, and Kristine were going to set up data loggers that would record soil moisture, temperature, and other parameters at preset time intervals through the summer. They were looking for a good place to set up their equipment. They were particularly interested in a site with obvious thawing permafrost (thermokarst).

The next three days Dana, Jack, and myself walked to sites and collected the data we needed. I think we were doing four to six sites a day. Mark and the others had not yet found the ideal site to set up their sensors and data loggers. They were also collecting some soil samples for microbe DNA analysis. Dana and I went down and used the chainsaw one evening to cut down some of the larger spruce trees down by the lake for a scientist in Fairbanks doing tree ring studies.

On about day four, it was our team's chance to use the inflatable kayak to sample sites on the far shores of the lake. The next day was rainy, and Dana and I were going to go with Kristine to sample a site up the lakeshore a bit. We piled in the boat. I had my GPS on its lanyard around my neck. The paddles were kayak paddles, and every time I took a stroke, my GPS slammed into the paddle section crossing in front of my chest. I was in back, so doing most of the steering. I knew this GPS and paddle thing was not going to work, so I hurriedly pulled the lanyard off over my head and tossed the GPS to Kristine who was in the middle of the boat, facing me. She said, "No, don't...," but the GPS was already in the air.

Kristine was not a particularly good catcher, it turned out. I watched helplessly as it slipped between her hands, bounced around and went overboard into the water. Dana, who was paddling in front, turned around in time to see the GPS splash into the water. I probed the full length of the kayak paddle and didn't touch bottom. It was then that I realized I had no other record of the detailed locations of the sites we had sampled. I only saved those locations on the GPS. We all were in shock, particularly Dana and myself, as we knew all the data we had collected was of little use for remote sensing mapping unless we had those coordinates. We proceeded ahead to the site as Kristine and the others had sampled in this location before and flagged it. It was a miserable site to sample, since not only was it raining with us sampling in the mud and wet ash of the fire, but also the dreaded problem of the lost GPS and its coordinates dominated my thoughts. What to do?

As we paddled dismally back to camp in the rain, Dana and Kristine suggested I ask Mark if I could use his GPS, since I think another on Mark's crew, Kim, may have had and another GPS. Back at camp, I reluctantly asked Mark if I could use his GPS. I explained that I had printed out the potential sites that were in my GPS and could enter them manually into his GPS. This would let us get back to the vicinity of all of our sampled sites. We had adjusted each site's center location to better ensure that 90 by 90 m area was all or mostly in the same site and land cover types. Guardedly, Mark agreed after some pleading from Kristine. I spent the next couple of hours meticulously typing all the coordinates into Mark's GPS and double checking them to be sure they were correct.

The next morning, Dana, Jack, and I all headed out into the dripping wet forest, retracing our route for the last three days as best I could. I knew which potential sites I was heading for when we initially approached each site. Once I got close, I announced to Dana and Jack that it was "somewhere around here." We had not marked our sites with flagging because we did not intend on ever returning. But Jack had taken some pictures of the center of the site with his camera where the GPS had been averaging and where Dana had done her biomass clippings. So, we knew the shape and configuration of trees we were looking for, and we looked for the bare spots where Dana had clipped. We looked and looked at the first site. I was about to give up when Dana found her clipped plots. From there, with Jack's pictures, we were able to get to where we were confident the center was. I took a quick GPS-averaged position and we were off to the next site.

Some sites we searched for hard and long, but ultimately found where the center was. One was a particularly spectacular find. I was coming at a site from the direction we had left the site, complicating my approximation of our previous route. We had to go through what appeared to be an old burn with really dense black spruce forest, with the trees just tall enough so you could not get your bearings. We essentially were swimming through the forest using our arms both to move branches and to provide some locomotion. Glasses or goggles were a must! I was about ready to concede defeat on this last site when I popped out into an opening. I looked around and things looked very familiar. When I looked down, there on the ground to my left was a bit of flagging Dana had left at the site center. Dana popped out behind me and I just pointed and she said, "No Way! Did you come out right here?" We could hear Jack battling the dense trees. He yelled that he did not think we would find the site. I yelled back, "Yeah, you are probably right. Come on up here, let's take a break." Three minutes later Jack popped out into the little clearing. He looked around and casually said, "You found it." I got the GPS coordinates (we were now also writing them down, per Dana's insistence). All that was left was to get back to camp! That walk back to camp was easy since we were all on cloud nine. We strolled into camp with smiles on our faces. Someone at camp asked if we found the sites and I said, "Every single one. Fifteen consecutive needles in a haystack!" I think this even might have reached legendary status in the folklore of the University of Alaska, Fairbanks, and the USGS Alaska field crews, as over the following years, when I shared this story with a new person on the crew, they had already heard about it.

Mark, Kim, and Kristine left the next day (I kept Mark's GPS) for Canvasback Lake. They never did find a good site to set up their instruments at Boot Lake. Jim, the bush pilot, picked the three of them up in his float plane, but due to a schedule issue, Jim could not come back to pick Jack, Dana, and myself up. Mark was going to get on his satellite phone to see if he could get another bush pilot to pick us up and take us to Canvasback Lake.

I remember it was an idle day, as we had to stay close to camp, not knowing when (or if) a plane was coming. I think Jack was doing some work preparing some of the soil DNA samples. As I recall, Mark even took Dana's satellite phone with him, just to be sure he had enough call time to negotiate a bush pilot to pick us up. As a result, we were there with no communication capability if something went wrong. A precarious position to be in. About 1 p.m. we heard a plane. A blue and white plane taxied up to our landing on the shore before I could get there. This pilot was Andy, and there was some issue with the number of hours he could fly per day. He was going to be right up against his daily limit. He wanted us to load our stuff as soon as we could. Jack had anticipated no flight to Canvasback Lake until the following day, but Andy made it clear it was this or nothing. We were breaking camp fast, just throwing stuff together. I was assigned to get the kitchen tarp down because I got my stuff packed first. Jack insisted that I not cut the cord tying it to the trees. I shimmied up the trees but found the knots were granny knots that had been pulled tight in the rain. No way was I going to be able to pick that apart with one hand while climbing a tree with my other hand. I whipped out my Swiss Army Knife and cut the cord. I quickly wrapped up the tarp.

When I got down to the plane, Jack was still packing his stuff. He asked about the cord and the knots. I told him I had to cut the cords as time was running out and they were not quick-release knots. We climbed in and the plane took off. We could see one of the burned hillslopes near the lake. The sun angle was just right and we could see sun-glint from every little rivulet of water seeping across the hillside. It looked like at least 50 percent of the hill had water flowing over it. Over the years, my theory for the abundance of water after a fire was: 1) the melting of near-surface permafrost or seasonal ice caused by the thinner, insulating organic layer and a darker soil surface after the fire, and 2) the complete reduction of plant transpiration (water that plants lose to the air during photosynthesis) because all the plants were dead. Transpiration essentially sucks water out of the soil and pumps it into the atmosphere.

With Andy, we flew northwest out of the loess hills where Boot Lake was and across a very broad flat plain, the "flats" of Yukon Flats. This plain is believed to be made up of old fossil river channels of the Yukon River as it migrated north to its current position along the northern edge of Yukon Flats. Andy touched down on Canvasback Lake, and we taxied up to a small dock. Mark, Kim, and Kristine helped secure the plane and unload our gear. Lickety-split, Andy was up and away, hoping to get back to Fairbanks before his flight hour limit for the day ran out.

There was a cabin at Canvasback Lake, but our take on a walk through the cabin was that it was crowded and nearly all the bunks claimed. Dana, Jack, and I opted to sleep outside in our tents. We sampled hard and got lots of sites our first sampling day at Canvasback. Dana had a fever one day when we had to cross an old burn with logs suspended at chest and waist high. We were constantly climbing up, over, and down burned log piles. To top it off, it was hot and there was drying fireweed everywhere. We were constantly inhaling the fluffy fireweed seeds, which billowed up as we walked. Dana, being much shorter, had an even more difficult time on that traverse. After that day, Dana said she was done. Jack said if she was done, we all were done. We could not continue without Dana, as she was critical for the vegetation composition and species identification.

The next day she was feeling better, and she agreed to give it a try. We walked slower and the routes were easier. We had discovered that the spruce trees we were seeing were all White Spruce, even though many had the typical growth form and growing conditions where we would expect Black Spruce. The only way to tell the difference between Black and White Spruce is that the new tiny twigs have tiny hairs on them if it is Black Spruce. Jack was up ahead scouting for some Black Spruce. He had dropped his pack while Dana rested, and I stayed with her. I noticed a cow moose rapidly trotting across the open grassy lowland toward Dana and me. The moose kept looking back behind her, roughly in the direction where we knew Kristine was busy installing arrays of soil sensors and data loggers. I mentioned to Dana, who was laying in the grass, that the moose was headed our way. She got up on her knees, and when she saw the bearing the moose was following and its speed, she became concerned. She jumped up and started yelling and waving saying that we did NOT want a moose coming toward us!! I had my camera out and was shooting pictures and waving with my other hand. My yelling was half-hearted compared to Dana's. I have to admit, I was kind of hoping for a better picture of the moose (not smart!). Finally, the moose saw us and stopped in confusion. But the new bearing the moose had taken headed right toward where Jack was looking for some Black Spruce! Soon the moose spotted Jack and changed her bearing yet again. This poor moose just ran into three groups of people within a half a mile, and I am sure she was completely surprised and baffled by all these humans out in the middle of the wilderness.

When out in the Alaska bush, it is always prudent to have bear pepper spray. One never knows when you will run into a bear, or vice versa. Kristine in our group was certified for firearm defense against wild animal attacks. Since Dana, Jack, and I were often not anywhere near where Kristine was working, we relied on the pepper spray and kept it handy.

Dana and I wanted to start working our way toward Jack, so I was able to temporarily strap Jack's pack on top of and on back of my pack. It was a heavy load, and I was pretty unstable, but Dana and I just proceeded at a slow pace. When we met up with Jack, he reported no Black Spruce, only White Spruce. Yukon Flats was an anomaly because of this lack of Black Spruce, with White Spruce filling the ecological niches typically occupied by Black Spruce. We proceeded to my next GPS point. Finally, the GPS indicated we were within 5 meters or so, and I announced that the site was somewhere near here. We all started to shed our packs to scout out the best position for our sampling protocols. I had just lowered my pack to the ground when I heard a loud "PSSSST," at the same time that I felt a blast of warm air on my thigh. As my head swung around, I saw Jack with his pack halfway down his thigh and a brown cloud waist high between Jack and me. I thought to myself, "That pepper spray does not smell too...", and then I could not breathe. The old firefighter instinct was to stoop closer to the ground (where the good air is in a fire) and run upwind. Jack ran downwind. Dana was laughing her head off, as she was safely out of range. Jack and I were cracking jokes when we could get words out between coughs. In hindsight, when I ducked down in my exit, I ducked into the low cloud of pepper spray. Jack was using his water bottle to wash his face, but I only had half a water bottle left, and I wanted to save it for the hot walk back to the cabin. We sampled that site, telling jokes and teasing each other. It was an open, scattered shrubland with a few scattered trees, so sampling went fairly fast. I am not sure the cause of this accidental pepper spray incident, but I think when Jack was out scouting for Black Spruce, brush or tree branches must have knocked the plastic safety clip off his pepper spray. Jack carried his pepper spray on the side of his cargo pants leg, in a tall skinny pocket that I think is for a hammer handle. When he took his pack off, and I can attest that it was heavy, he probably moved the pack to one hip and slid it down his leg, impacting the pepper spray trigger.

We slung our packs on our backs and started the long, hot walk back to the cabin. I was leading, per usual, but I soon noticed that there was a slight headwind, and Dana and Jack behind me were coughing in my pepper spray fumes. I stepped aside and explained that I would bring up the rear as long as the wind was from this direction. It was hot and painful, as where my hat connected to my head and where my shirtsleeve cuffs rubbed my wrists were particularly irritated by the pepper spray. Finally, we walked up to the cabin. Dana and Jack, being in the lead, started briefing the others on what had happened. I stalled, planning how I would wash this stuff off and change my clothes. As I was digging around in my tent, Dana asked to borrow some liquid soap I had and I complied. Soon I had the clean(er?) clothes ready, and seeing Dana coming back from the lake, got my soap back. I trudged down to the dock, dreading what was about to come. I knelt down on my knees on the dock, and with the soap open and handy, I closed my eyes and kept them closed as I rinsed my face and arms in lake water, then lathered with soap, then rinsed, then soap, then rinse, and a third round of soap and rinse. I grabbed a tee shirt that I was using as a towel and dried off. Then I opened my eyes and boy did it burn!! I think I was moaning or even shouting in pain. Then I remembered that milk was an antidote for spicy food. I had seen a box of dried milk in the cabin. I yelled to Jack, who was in the cabin, to make me some milk. His reply was that the active ingredient in milk was the oil and the dried milk in the cabin was fat free skimmed milk. He assured me that the pain would subside in 5 minutes or so. I guess that comforted me as I grimaced in pain a bit. Finally, the burning started to subside. I removed my brown pants with an orange-brown stain on one thigh and double bagged them in plastic bags.

Jim Webster was going to make three trips in his plane to get us home. The question was, who was going home on which plane load? I wanted to stay and go back on the last plane. I liked it out here in the wilderness, pepper spray and all. I wanted to be the first in and last out (a touch of machoism, but I did really like the wilderness). Kim took the first plane home, taking a lot of gear with her. She did ask Dana, Jack, and myself what we would like to drink the most when we got back to Fairbanks. I said a beer and Dana said wine. Webster's plane returned the next day to get Mark and Kristine and more gear, but by then Jack, Dana, and myself were off sampling again. When we returned to the cabin, we found a tall can of beer and a box of wine! We sat out and watched the moon rise over Canvasback Lake as we shared our adult beverages around the picnic table that was there. We flew out the next morning, but Andy, in his blue and white plane, after hearing about my pepper spray incident, put my double bagged pants where hazardous cargo goes, inside the plane floats. That way the pants could not affect us, and particularly Andy, the pilot, in the plane cabin.

I think I passed all my Alaska bush initiation. I am certain the pepper spray incident was not a planned part of that initiation. I returned every year after that for field sampling, but focus drifted to biomass and then to permafrost, as those map products generated the most interest. Lots of field sampling trips in the interior and several to the North Slope. I think both the pepper spray and the GPS in the lake stories were repeated more than once, as often new members to our field crew in subsequent years seemed to have heard those stories before and responded with, "Oh, I heard about that before. Was that you?"

BEAR IN CAMP (2010)

It was mid-June, and we were headed back to Yukon Flats to get more permafrost, biomass, soil moisture, and vegetation composition data for mapping. A work colleague at EROS, Kurtis, had a strong interest in both Alaska and remote field work. Kurtis was interested in the canopy structure of the vegetation, as he was working on providing that type of information in maps that could better inform wildfire behavior. I had Kurtis take the necessary float plane and bear firearm trainings need for the trip. Kurtis was also an Emergency Medical Technician (EMT), so we had the first aid covered as well. Dana, up at Fairbanks, was also coming along with a fellow University of Alaska graduate student, Mark. Mark was a big, strong guy with a beard and dreadlocks, who knew his plants and Alaska ecology. I had agreed to give Dana and Mark one day out of the ten days in the field to get data for Mark's PhD dissertation. This was the only field trip I did in June. It was hot, lots of bugs, and it did not get dark until midnight (tough to sleep in a tent)! I preferred the fall, as the bugs were knocked back by early frosts, we could be at the maximum thaw depths with our permafrost probe, and the fall colors are simply stunning up there with the birch and aspen turning bright yellow and the tundra vegetation often turning red.

Jim Webster again was our bush pilot with his brown and white float plane. I had sent him imagery of the lake I wanted to go to, the distance it was from Fairbanks, and the length and width of the lake (takeoff and landing distances). This particular lake was kidney-shaped and was of interest because it was at the base of the loess hills and the southern edge of the Yukon Flats plain. It was north-northwest of Boot Lake. Our preliminary remote sensing estimates of biomass from our preliminary biomass map indicated that this was an especially productive area. Another analysis I had done showed that this area was significantly more productive than a Black Spruce forest. I wanted to validate what we were seeing in these analyses and hoped to better understand the causes of the peculiarity of this area.

Kurtis and I were on the first flight in. After flying over the lake, Jim said it was OK for landing and takeoff and asked if we had picked a campsite. Based on the GPS sites I had picked, the southeast edge of the lake would be less walking. The landing was smooth, and we taxied up to the shore, killed the engine, and coasted in. We strapped on our hip waders, as the lake had a mucky bottom that would only let you remove your foot from the muck with a lot of effort. The grassy marsh on the edge of the lake was no cakewalk either. It was kind of a floating mat that moved as we hauled gear to firm ground. Our path soon started to sprout holes where one could end up sinking to above the knee. After a fair amount of hard work unloading, Jim took off to get Dana and Mark back in Fairbanks. Kurtis went to scout for a campsite while I scavenged logs, sticks, and brush to lay on our marsh path so that that Dana, Mark, and the rest of the gear could be hauled to dry ground with a bit less effort.

The lake kind of smelled like dirty socks and looked like strong tea. We were bringing Fairbanks water that would last us five days, but we would be staying 10 days. So, the brutal truth was at some point we would be dinking this lake water. We had a filter, but I was hoping the filtered water would not taste and smell as bad as the lake water.

Kurtis and I started hauling gear up to the camp area Kurtis had selected. When we heard Jim's plane approaching, we got down to the lakeshore in time to watch the landing. Dana and Mark had brought small, light inflatable kayaks to facilitate getting around the lake. With four of us in waders, unloading went fast. I asked Jim if, in his 30+ years of flying in Alaska, he had ever been to this lake? "No," was his response. Jim said his good-byes and taxied to the downwind edge of the lake, gunned the engine, and lifted off. When that bush plane left, the realization that "we are not in Kansas anymore" hit me. We had a satellite phone for daily check-ins and to report weather conditions to the pilot on the morning we were to get picked up. But, any injury or major problem that developed, we pretty much would just have to solve it ourselves. This makes me conservative and careful. It's easier to avoid problems than solve them. I gave a very short safety reminder to everyone as a reminder, but the importance of avoiding problems was obvious to us all.

Our tents were in a string heading away from the kitchen area. We had a screen tent (thankfully) where we could at least get a break from the mosquitos and flies. We took some time getting our cooler suspended high up between two trees to discourage bears. Mark discovered one of the kayak paddles had broken in half. He planned to hold each short paddle in each hand to paddle. I said no, just splint it. Mark did not think that was possible. I grabbed a birch branch and some parachute cord and splinted the paddle back together. Mark said it worked fine. Our first meal was salmon (caught previously and provided by Dana) down by the lake. We dug a small fire pit and watched it closely so it did not start creeping into the duff layer and spread. With a little wine, the crew's spirits were high.

The first "shake down" site was close to camp; that way, if you forget something, you can go get it. Not 100 yards from camp, Dana started swearing. She had just got stung by several wasps. She looked around and there was a big paper wasp nest right next to her. She was the third person in our hiking formation that day. Three of us had walked right by it, startling the wasps, so apparently, they were ready and waiting for Dana when she sauntered by. This was a major concern as Dana had told the group that she was allergic to bee stings. I told her to take some Benadryl. She took her pack off and dug around in it before stating that it was not in her pack. She thought she must have left it in her other pack, which was back home in Fairbanks. So, I turned to Kurtis, the EMT, and asked him to give her some. His reply stunned me. Legally he could not administer drugs!! I checked my pack, but no Benadryl or antihistamines. All we could do was sit and wait to see how Dana reacted to the stings. Fortunately, nothing happened, but I was frustrated that all I could do was just sit there and wring my hands. Kurtis could have done a tracheotomy if needed, but that would only have been under a life or death situation.

The first site had large scattered willows. Some of the basal diameters on these were basketball sized (the largest Dana had seen in Alaska). There was evidence of the overland flow of water, but I never did figure out where the water would have come from. I suspect snowmelt of the loess plateau to our south. As we pushed on to our next site, we encountered dense shrubs in the understory, making walking very difficult. We scratched our way around for two days, sampling four to five sites per day. We had sites right at the base of the loess plateau and up on top. On the day we were heading up top, Kurtis noticed a water bottle of his was missing, again within 100 yards of camp. I went back and helped him look for it. After a while, he returned to his pack and got ready to go. I assumed that he had found his missing water bottle. The hike up was not as brushy, but climbing the 50-foot steep bank to the top made our hearts pump hard. Once on top, we sampled a site and I got stung about seven times by wasps. This time the wasps' nest was below ground. No big deal, worse than a mosquito bite, but I pretty much forgot about it in an hour.

My GPS showed the next site was quite close, but it turned out we had to drop down and cross a deep little canyon with a trickling brook at the bottom. So, we had to climb the steep loess plateau yet again! It was getting hot, and now we were walking through an old burn. This burn was recent, the same fire that was actively burning when we were at Boot Lake the year before. The fire extended right up to the edge of the loess plateau before going out. Hiking and sampling was dusty from the ash but also significantly warmer as there was little shade (just charred, lifeless trees), and the dark ash was absorbing the sun's heat like crazy! We had two or three sites in the burn that went a bit slow, primarily because Mark and Dana were finding Morell mushrooms everywhere. They collected the mushrooms and took them back to camp where we used some in our meals, but most were left to dry in our screen tent. This made for some interesting smells for the next couple of days, which bothered Kurtis a bit. After finishing another burned site, we took a lunch break. It was then that I noticed Kurtis only making a feeble effort and not drinking copious amounts of water like I was. It turned out that he had never found his second water bottle and depleted his other one. I dug around in my pack and came up with my spare water bottle and gave it to him. I told him that he should have said something! I said that I did not want a martyr! "Let's work together and get back to camp." On the way back to camp, Dana also shared her water with him.

Kurtis was acting strange, a bit logy and lethargic. Alarm bells went off in my head when he did not eat much at supper. Finally, he acknowledged that he was not feeling well. I encouraged him to drink lots of water and get a good night's sleep. But when morning came, he announced at breakfast that he was going to be staying in camp, as he still did not feel well. Mark had gone off into the woods after breakfast to "visit mother nature." When he came back he asked, "Guess what I found?" I had no idea, but Dana quickly put two and two together and she said, "Kurtis' water bottle!" and she was right. I suggested that on this day, Mark and Dana go get data he would need as part of his dissertation. This was their day of sampling that Mark and I had agreed to earlier. I was going to stay in camp and keep an eye on Kurtis and start filtering lake water to fill some of our kitchen water jugs. Dana and Mark busted down to the lake to inflate their kayaks and test the splinted paddle.

Kurtis and I did one site just across the lake, but I could see he was weak, so we headed back to camp after finishing one site. We were lounging in the screen tent. Kurtis had used up all my Airborne effervescent tablets. Kurtis said they made the stinky lake water more bearable to drink (lemon flavoring plus the carbon dioxide bubble may have aerated the water a bit). Then Kurtis thought he heard something, turned, and exclaimed, "A bear!" He said it in an "excited to see you," happy voice. I jumped up and there was a black bear walking up our path from the lake. I immediately felt trapped in the screen tent and unzipped the door. I stepped out and grabbed my shotgun, which was nearby. Kurtis followed and grabbed his camera. He took a couple of pictures while I started yelling at the bear and waving my hands. Later Kurtis said at this point he realized grabbing a weapon (a club or pots and pans to bang?) might have been more prudent than the camera. My vocalization and gyrations did not scare the bear at all, so I yelled louder, using obscenities with more feeling (that should help!). Kurtis was also yelling and waving by now, but the bear just sat there, 20 yards or less away, peering through a maze of willow branches (most of the willow leaves were higher up) at us.

I decided the situation was changing to herding the bear out of camp. My gun was always in "field ready" condition—that is, the chamber is empty, safety is on, and the magazine is full of slug rounds (the bullets were about as big as my thumb). So, all I had to do was slide the fore stalk back enough to get the bean bag round—two of them were in two little elastic sleeves on the rifle sling meant for holding additional rounds—into the chamber. I wanted to thump the bear with my bean bag round to punish him and scare him off with the sound. I moved forward down the trail trying to act confident, shotgun in firing position, and yelling as I advanced. I was looking for a gap in the willows where I could get a shot off. Kurtis said later that he was a bit startled, as he thought I was going to charge the bear or something. As I got closer, the bear got nervous and ran across in front of me, 15 yards away. I had my sights on him, tracking him, waiting for a gap in the willows. A gap appeared just for an instant, and I fired. The bear continued running down the trail, back down toward the lake. Kurtis and I went to investigate, and Kurtis was surprised that I was able to get a shot off in all those willows. I found the window I shot through. Kurtis walked down the lake trail so we could triangulate where to do a detailed search for the bean bag. Kurtis found the bean bag and a spot on a large willow branch were the bark had been blown off by the bean bag's impact. I had missed! Just a half an inch high. Kurtis and I walked down toward the lake and saw something black run away through the thick brush. Kurtis wanted to follow it, but I reminded him that we were not hunting, only here for self-defense. We walked on down to where we unloaded from the plane. I had left my hip waders down there, as I saw no need for them up at camp. The bear had dragged them around and taken a big bite out of the left knee. I spent that evening duct taping my hip wader back together. How else was I going to load gear and get in the float plane for the trip home? Kurtis cooked supper, tacos. Kurtis hated doing dishes while I hated cooking. I was a bit worried about Dana and Mark still being gone, but had confidence in them. Finally, about 10 p.m., it was still pretty light, I took my shotgun and went to bed. Kurtis was going to bed as well, so we left them a note about where the leftovers were and to keep an eye out for the bear I had shot at.

Dana and Mark did not return until something like 1 a.m., but they saw our note and had seen my hip waders with a bite out of the knee down by the lake. While eating the leftovers in the screen tent, Mark noticed Kurtis's camera and checked the recent pictures. There was a picture of Mr. Bear peering through a maze of willows. Mark said to Dana, "This tree branch in front of the bear on the picture, it is that branch right there." He pointed to the tree we had holding a water sack with a nozzle for washing dishes and hands. Having seen that the bear had actually been up near our kitchen and sleeping area, they went back down to the lake and carried their inflated boats back up to the kitchen area. They were hoping that they would at least hear a bear in camp before both of the kayaks got shredded, as long as at least one of us was a light sleeper.

The next morning, I was up before the others and got the coffee going as usual. Soon Kurtis piled out of his tent. We were running low on coffee, and it looked like the last morning was going to be with no coffee, a cardinal sin in camping. Dana and Mark crawled out of their tents and shared their story about their late arrival back at camp. Both Kurtis and I had heard them come in last night, having our bear awareness turned up a couple of notches. But Mark's sampling had not gone as planned. Dana and Mark had spotted a bear circling them back in the woods as they collected the data Mark needed. They had Dana just stand guard with her pepper spray while Mark finished up a portion of his sampling. But given the persistent and curious bear, ultimately, they decided it prudent to abandon their site and return to camp. They asked if I could bring the shotgun and accompany them back to Mark's site to stand guard while they finished Mark's data collection. I looked at my GPS and noticed I had two possible sites just north of our lake while Mark's site was northwest of the lake. Kurtis was still not feeling his oats and still felt weak. I proposed that Dana, Mark, Betsy (my shotgun), and I go north and sample two of my sites, then proceed on to Mark's site. This would give Kurtis another recovery day in camp. While sampling one of my sites, I asked Mark how he knew that these spruce were Black Spruce. I asked because the trees were fairly close to each other and tall, with few branches within reach. He pointed out the tall spindly shape of the canopy and the cluster of pine cones at the very top of the tree. Then to verify his observation further, he walked over to the only branch within reach, reached up on tiptoe, and broke off some new twigs for inspection. To both his and Dana's surprise, they were White Spruce!

After finishing up two sites, Mark led us to his site. Mark's GPS got us close, and then they started remembering where they had sampled. As they began to unload their notebooks and sampling equipment, I noticed something strange on the ground, a long, skinny stick of wood with no bark, and it seemed to have grown in spurts with distinct bands around it. Dana said, "Oh, that is the tree core I took yesterday to get the age of the trees. And you are standing on a wasp nest because that is why I had to leave that core." Suddenly, right on cue, there were several wasps swarming around my face. I swatted them away as I moved quickly to another location. I just kept my eye out into the forest while they wrapped up their data collection, with Betsy slung over my shoulder. They both thanked me and said it was very reassuring to have Betsy and me along.

We headed back toward the lake and encountered a smaller lake where we got a water sample that another group of hydrologists had asked us to collect. A pair of Sandhill Cranes were in the tall grass on the other side of this small lake watching us and calling (sounding the alarm bell?). I challenged Mark and Dana to a crane calling contest. Sandhill Cranes have a peculiar, almost prehistoric call. I went first and produced nothing that even closely approximated a Sandhill Crane, but did get spittle all over my chin. Mark was next and did better than I, roughly mimicking the Crane's call. Dana, however, produced a very close call to that of the Crane's, and they rewarded her with a call back in response.

It was finally the day for the float plane to pick us up. It would not be Jim Webster, but the Fish and Wildlife plane with a pilot named Mike, who I had met once before. We had called Fish and Wildlife our first morning to provide the GPS coordinates of our camp to aid Mike when he came to pick us up. We were up early, as you never know what time the plane will come, and we had a lot of gear to get down by the lake. To exponentially increase the difficulty of our departure, we were out of coffee. It was a quiet breakfast with little conversation. I took the first load of gear down to the lake, leaving Betsy in camp with the others. Having gotten up first, my tent and gear were ready before the others. I took a second load of kitchen stuff as others made cargo hauling trips. Soon, I needed a break. I sat down on the ground a yard from were Betsy lay, looking out into the forest. The others started retuning from the lake, and Mark came over by me as I drank stinky, dark water from my water bottle. It kind of looked like coffee, but definitely did not taste or smell like coffee. All of a sudden, Mark calmly said, "There is a bear right there." I could see he was pointing to the willow only 7 yards away. I stood up and grabbed Betsy in one motion. I could hear branches breaking as something ran off. We could hear it circling back toward the lake as I moved to position myself on the edge of a small opening in the trees so I might see the bear and be able to shoot if needed. All I saw was the bear for a sixteenth of a second as it sped toward the lake. Mark later told us the bear was just hiding under the willow watching us. I was both surprised and mad. How had I and my crew not seen it? A bear that close is a big safety concern, and I wanted no such mishaps on this or any other of our field sampling trips.

We finally got all the gear down to the lake and decided to leave the bright red and bright blue kayaks inflated so the pilot could see us better. As we lounged by the lake, I kept an eye and an ear to the forest behind us. I used a timed delay on my camera to get a group picture. Finally, we heard a plane and watched it fly around the lake many times. It flew right over our campsite once, very close to us sitting on the edge of the forest. The plane rocked its wings and came in for a landing. I commented that I had never seen a pilot check out the landing that much before. We learned later that Mike, the pilot, had not seen us. Mike told us we should be more active when the pilot is looking for you.

Mike landing at Hornet Lake to take us to Fairbanks.

In our lake lounging time, we hatched a plan to load the plane easier. We had attached long parachute cord to both the stern and bow of one kayak. This way we could ferry our gear to the plane in the kayak with us on shore with one cord and pilot with the other cord. Kurtis used the other kayak to take a load out to the plane and deliver the tow rope for the other kayak loaded with gear. Somehow, when near the mucky shore, Kurtis lost his wedding ring. He looked for it in the 6-8 inches of muck, feeling with his fingers, but ultimately his ring remains back in that lake. Dana suggested that I should name this lake (for our data sheets and for possible future visits), and the others concurred. I wondered if I should name it "Ring Lake." Or how about "Kidney Lake," since it was shaped like a kidney? Ultimately, I announced that it would be called "Hornet Lake."

Mark and Dana were the first planeload back to Fairbanks, and they told Mike all of the bear stories. Kurtis and I were the last out, and when we got to Fairbanks, both Dana and Mark were long gone. Kurtis and I photocopied our data sheets, a copy for me and a copy for him, so if one of us lost our luggage, we would still have the data. We had equipment to return and lots of errands before flying home from Fairbanks.

— 29 —

GRIZ (2013)

We had made permafrost maps of Yukon Flats, then the Yukon Basin in Alaska, and now we were looking to extend our mapping to the North Slope (see http://dx.doi.org/10.1016/j.rse.2015.07.019). For mapping such a large and remote an area as the North Slope, there was a distinct lack of a good geographic distribution of permafrost field observation from maximum thaw time. Most of the existing field data we found was concentrated along the only road through the area, the Dalton Highway, or "Haul Road," and near the town of Barrow. This time I was traveling with a former summer student who we kept on because of his productivity. Neal was in his early twenties, well over six feet tall, athletic, and the face of a high school student (Neal always gets carded). Neal and I headed up the Haul Road, sampling as we headed north, and camping on some short pull-out roads off the main road. We had some trouble with our 110 AC adapter, which we needed to use Neal's laptop to view satellite images and aerial photos in our rental truck. It was a productive trip, with only one broken shovel, thankfully at one of our last sites. We met a bicyclist who had flown into Barrow and was riding down the Dalton Highway to at least Fairbanks. We passed him multiple times as he would pedal past us when we were out looking for a site or sampling a site. He had his ear buds in and paid no attention to the heavy truck traffic on the road. He just focused on pedaling. Neal and I arrived at the empty campground near the Arctic Circle sign. We picked a campsite close to the restroom, but not too close. I had my shotgun leaning against the picnic table as Neal and I set up our tents and collected some firewood. Then two cars full of Asian folk pulled in right across the road from us. We were kind of wondering why they wanted to be so close. I wondered if they saw the shotgun and thought it would be best to be close to that, or was it the restroom?

As Neal cooked our supper and we dined, the bicyclist pedaled in, circling the campground roads and checking out the campsite options. He picked one next to the thick brush on the eastern edge of the campground. Neal and I had been up early that morning and had done a lot of walking, hole digging, and probing the soil for permafrost, so we were not going to be staying up late. I was awakened in my tent by the slamming of a car door. I dozed off and then heard another slamming car door. I could hear the click and whining of a camera timed exposure being used to get Aurora pictures. "Click, hirr, hirr, hirr.........SLAM!......... click, hirr, hirr, hirr." I stuck my head out my tent door to only see mediocre northern lights and to shine my flashlight off in the direction of the car with the slamming door issue. After multiple repetitions of this, I resigned myself to the situation and just tried to ignore it as best I could, vowing to get up early and make some noise getting breakfast and our sampling gear in order.

Morning came after a fairly sleepless night, and Neal, too, acknowledged the noisy night. We ate and broke camp in fairly short order, having already done it numerous times on this trip. I did try a scowling glare at a couple of Asian folk headed to the restroom before realizing that the elderly woman was somehow handicapped with a severe limp. Neal and I took off fairly quickly after that.

A couple of days later we were approaching the Yukon River, with the "Hot Spot" camp and the "Yukon River Camp" being about 7 miles apart. We stopped at the "Hot Spot" and got a burger, fries, and a milkshake, which all tasted divine after eating our camp food for a week. Neal was a good cook, but our fresh meat and vegetables had long since disappeared. I was the dishwasher, but not terribly good at it or proud of it. As we were finishing, the bicyclist pedaled up. He ordered some food and asked how many hills there were from Hot Spot to Yukon River Camp.

He was going to spend a couple of days recovering at Yukon River Camp. I reflected and I thought there were 2 or 3, but later as we drove to Yukon River Camp there were more like 10! We mentioned how we had been seeing him on the road and assumed that he would recognize our truck after passing him some 30 or more times. He was completely oblivious that we had even passed him once. He was a safety officer for some company in Vancouver, Canada. We mentioned seeing him at the Arctic Circle campground and told him we were the guys camped near the restroom. Then he remembered seeing us there. He said he was sleeping in his tent at Arctic Circle campground when he heard something in the thick brush behind his campsite. The noise persisted and he became quite scared as he was sure it was a bear. His solution? Put on his ear buds and crank the music up. I rolled my eyes in pain, clearly visible to Neal (who smiled back) and the bicyclist (who just shrugged it off). The bicyclist told of how once he was approaching a crew working on road repairs and they were all staring at him and laughing. Finally, they started pointing at something behind him, and there was a curious caribou following him down the Dalton Highway. He suggested that some of the caribou hunters we had seen along the road should try drawing a caribou in close with a bicycle. When we left him at Hot Spot and headed south, we never saw him again. Toward the end of our conversation, he said he was thinking of giving up his trip at Yukon River Camp and catching the weekly bus home to Fairbanks. He was going to rest at Yukon River Camp for several days, contemplating his decision.

Neal and I got back to Fairbanks and immersed ourselves in good food, drink, and sleep. But we had to prepare for our next outing, which was a 10-day float from the Brooks Range to the Arctic Ocean. We would be traveling with a permafrost expert who lived in Fairbanks, Torre. Torre was a tall, lanky, blond haired guy, probably in his late 50s, with an infectious smile. We heard from Torre that the Fish and Wildlife Service (FWS) strongly recommended that we not go down the Jago River as we had planned. The Jago was running high and the rapids were very tough. The FWS recommended that we float the Hula Hula River, which had milder rapids and was farther west. This put us in scramble mode, as we had picked all our spike camp locations on the Jago River. Neal worked on his laptop one evening, picking prospective camp sites and sample locations on the Hula Hula. The Hula Hula River would require the best location (GPS coordinates provided by Neal) to do a short portage over to the Okpilak River, which would let us avoid the big sandy and shallow delta of the Hula Hula at the Arctic Ocean. I had reserved an inflatable rubber boat, which was kind of a hybrid between a rubber raft and inflatable canoe. It had oars and one canoe paddle. We picked it up in the suburb town of North Pole, which was just southeast of Fairbanks.

Next was to link up with Torre. Torre had been studying permafrost in Alaska for many years and had numerous publications on permafrost. Torre was kind of our guide as well because he had done a fair amount of really remote field work and boating. Torre's daughter, Karen, would be going along as his technician. Torre planned our menus right there around his kitchen table with no notes. We were scribbling notes as we had a LONG shopping list. Torre, Neal, and I met at Fredd Meyers, a big chain grocery store, and each of us took a portion of the shopping list. Either I got the tough items, or I was much less efficient than Torre and Neal, being in an unfamiliar store buying many unfamiliar items. Toward the end, both Neal and Torre helped my get my remaining items. Two shopping carts mounted high were run through the cashier's scanner. I was wondering where we were going to put all of this food, plus supplies, in both our boats (Torre was taking his inflatable canoe) and in the bush plane that would take us to a gravel bar on the Hula Hula River in a Brooks Range valley.

The next big task was packing the vehicle, a full-sized Suburban. We met at Torre's house again to load the Suburban. We had bear barrels (plastic and metal containers that are supposed to be bear proof) and all our personal gear. I knew we were way over our weight limit for the bush plane. "Z", the head of the USGS LandCarbon program, was going to go with us up to Cold Foot on the Dalton Highway to drive the Suburban back to Fairbanks. We all squeezed into the Suburban about 8 a.m. and headed north. We ate lunch on the banks of the Yukon River near Yukon River Camp. Then we pushed on to Cold Foot. Cold Foot has a truck stop with a hotel, an airport, a BLM visitor center, a phone, a truck repair shop, and a few cabins hidden in the woods. We drove over to the airport to meet our pilot and get any last-minute instructions. Dirk was a skinny, wiry guy with a friendly smile. He informed us that there had been some weather issues that had hampered his flying the last couple of days, but the forecast was good for tomorrow. He had a Dahl sheep hunter who had been waiting several days to get out to his hunting spot. Dirk told us to come around 8 the next morning and weigh our gear. We had hoped to get all our gear in one flight, but I was very skeptical.

The big draw of Cold Foot was the buffet (top notch), beer, and gas. We enjoyed a good meal at the buffet and then spent the night at several BLM cabins about 5 miles north of Cold Foot. The next morning was frosty! It was cold in Cold Foot? This gave me some concern. Here it was late August, and it was cold and it had just been raining for the last several days at Cold Foot. What were we getting into? At the airport, we got instructions from Dirk on how to weigh our gear. I mentioned to Torre that I thought we had too much gear and sure enough, we were just over the weight limit for one planeload. Dirk said he would take us and our personal gear and bring the boats on the next flight. Dirk asked for some billing information for the flight, which I provided him.

We did a lot of standing around after that, waiting for the sun to break through and for Dirk to get a weather report from some other pilots flying that day. We loaded our gear on a big wagon and dragged it over to a big blue and white Beaver airplane with tundra tires. The Beaver was huge compared to the Cessna float planes we flew in on Yukon Flats, so I was a little reassured that most of our gear would fit. Finally, with the sun breaking through the clouds, Dirk told us to climb aboard. We all looked at each other wondering who should go first, so I just turned and said, "I won't be the last one on board," and climbed in and scooted across the seat to the far window. Behind the seat our gear was stacked to the roof. All of us were in, and I was like a kid in a candy shop, checking out the plane. Neal was in the co-pilot seat up front. As we taxied to the end of the runway, Neal pointed out to me that the Dahl sheep hunter was also on board the plane. I thought that was weird, as in all my research field trips I had never shared a plane with anyone before. Dirk swung around at the end of the runway and gunned the engine. Soon we were airborne and climbing. It was then that we could see that the entire Brooks Range was covered in snow! Dirk was talking on the headphones that I had on about how this looked more like mid-September and that winter must be coming soon. Again, I thought to myself, "What are we getting into?"

It turned out that after we got back home, I found out that my billing information I had given Dirk was all bogus and from last year. One of the accountants from headquarters informed me that I had not followed the necessary protocols for funding a bush plane in Alaska. But, my financial accountant had done this for me on previous years and she had sent a request to headquarters asking for guidance on setting up the bush plane billing. No response came from headquarters, so she assumed that what we did on previous years would work. Unknown to us, the Alaska division had recently changed the bush plane protocols, but had not informed any of the other sections. The accountant gal at headquarters said that all of my crew should split the bill. I forget what the bill was, but it seemed like $20,000 or more, much more than I or any of our team could afford. Finally, a head accountant, Ann, sent an email saying that no, we would not be personally responsible for the bush plane flights, but said we would have to do a requisition in arrears. All in all, it took nearly 6 months before Dirk finally got his money for our flight.

Flying over the jagged and snow-covered Brooks Range was stunningly beautiful and pristine, but it also was a bit unnerving. The flight was long, and our extreme remoteness was constantly being reinforced. Dirk seemed particularly taken by the early snow, with his conversation returning to that theme. I would guess somewhere after the halfway point on our flight, I started to become aware of my bladder being rather full. My full bladder had the focus of my thoughts for the rest of our 3-hour flight. Finally, Dirk pointed out the Hula Hula River ahead. There were three small piper cub bush planes parked and several sets of tan, full-sized tents and a small, tall tent that looked like a porta potty. Dirk was on the radio announcing our intent to land as he made the downwind leg approach, surveying our landing strip, which was just a shrub-free stretch on a gravel bar next to the river. As we banked during our 180-degree turn to get lined up with the runway and heading into the wind, a guy climbed into one of the parked piper clubs, started it, and gunned the engine and rolled onto and down the runway. Dirk immediately was on the radio re-announcing his landing intent, specifying the piper cub take notice. The reply was only, "Give me a second and I'll be out of your way." Before the radio transmission was over, the piper cub was amazingly already airborne, climbing, and banking off toward the east.

We landed without further ado and rolled to the south end of the runway near the river. The Dahl sheep hunter camp was kind of mid-runway and west from the runway, and a fair distance from the river, which was just a bit east of the runway. Immediately, Dirk cut the engine and said we could get out. All the males headed off to the south and immediately began peeing. I am not sure what Karen was doing, but I was not going to look back. We stalled a bit, discussing the river and scenery and our plan to unload all our gear here and then drag it off east and camp in the scattered willows in the narrow stretch between the runway and the river. I asked Dirk if he was sure this was the right place, as it sure looked like the Himalayas to me. Dirk reassured me this was not the Himalayas and with a straight face invited me to look at the altimeter. Finally, one of us peaked back to the plane and started walking back toward it. This was the "all clear" signal concerning Karen and we headed back to unload the plane. In no time at all our gear was in a pile on the southwest end of the landing strip. Dirk said his goodbyes and said to call him with the weather report tomorrow. Torre had a satellite phone and agreed to give him a call. Dirk climbed in, started the engine, and taxied to the north end of the landing strip. We had wandered a bit off the landing strip as Dirk gunned the engine, raced down the gravel bar, and lifted off.

We busied ourselves moving our gear over to our campsite between the river and the runway and set up our tents. This was not what I had envisioned. I had signed papers with the FWS agreeing not to camp close to the river near this heavily used landing strip. But given our mountain of gear, I really did not see another realistic option. Our boats should be delivered by Dirk tomorrow, and we would be off downstream, headed north. Our plan was to do five spike camps on our way to the Arctic Ocean, so one day boating, one day sampling all the way to our rendezvous in the Arctic Ocean lagoon behind the barrier island, Avery Island, where Bruce from Kaktovik would be picking us up in his boat. There were a lot of sequentially dependent events that would have to occur for this plan to work. For instance, there was no wiggle room for bad weather or an injury or illness. But this mission was already in motion, so our focus would be on carrying it to a successful completion.

Our camp consisted of four small sleeping tents and one larger cooking dome tent. The next morning, we grabbed our sampling gear and headed out to go over our sampling protocols together with Torre and Karen. As we hiked out along the runway, a small red plane landed and two fish and wildlife folk walked over to us. They asked if we were the USGS team and we said "Yes." They had several questions for us. Karen had a sore toe and had a Bledsoe boot kind of thing on her foot. But Torre assured them that we did not discriminate against the handicapped. Finally, just before we parted ways, Torre explained that we were waiting for our boats to come in. We had planned on leaving this airstrip earlier, but with the boats on the second planeload, we were forced to wait at the runway. I was relieved Torre had brought that up so casually, as the FWS forms were very clear they did not want campers near the river near the landing strip. Out sampling, Torre had some great suggestions and guidance on data to be collected and things to look for. We did sites that day to ensure as much consistency between the two sampling teams, Karen and Torre on one team with Neal and myself on the other. Torre's motto on this trip was to get data from the talus slopes to the Arctic Ocean, so the next day Neal and I were headed up to a perched valley in the Brooks Range. Per usual Neal was way in front and my heart was pounding, and I was panting and sweating profusely. I am sure this was because I was carrying the heavy shotgun, but Neal may have been in a "tad" bit better shape than me.

Suddenly, Neal froze and said, "Bear! Get up here with that shotgun, she has a cub!" My reply was, "You come down and I'll meet you in the middle." From what I understood from Neal's description of what direction the bear had run, I suggested we side slope around to the southwest a bit and sample near a rocky outcrop over there. Neal agreed, but reluctantly. As we rounded the hill, there was the sow and cub, who halted, did a 180, and disappeared over the top of the ridge. I had misunderstood from Neal which way the bear had run! That rocky site, though, had a very nice view. Later back at camp, we learned from Torre and Karen that there had been a black wolf following us.

It was a cool night and we woke to about an inch of snow on our tents. This was a bit alarming, as from here we were headed north. I nervously remembered Dirk's chatter about an early winter. But there was not much we could do now. Neal and I headed off to the northwest sampling, while Karen and Torre headed southwest. Torre had called in with the weather report to Dirk. Dirk was really interested in how far Torre could see up the mountains. About mid-morning, when we were out sampling, we heard a plane and then spotted Dirk's blue and white beaver. Dirk landed and piled a bunch of gear by the airstrip and took off. I had a florescent orange stocking cap on, so Neal and I were very visible. As Dirk flew over us on his way back to Cold Foot, we waved and he wobbled his wings and disappeared. When I was trying to keep up with Neal, I tried to cut across a ridge covered in low shrubs, but instead the brush concealed a thigh deep, jagged rock-lined hole. I found myself suddenly sitting on the ground in the shrubs. I inspected my legs for injuries. Having found no blood, I was relieved, but I did find a big tear in my right calf of my waterproof pants. I was able to patch these up fairly well with some gorilla tape in my pack. A near miss though. No place for a broken leg or significant puncture wound.

We returned to camp and someone had heard that the Dahl sheep hunters would be leaving as soon as the weather lifted enough to get them out. We had a one-gallon can of white gas for our stove, but Torre thought it would be nice to see if the Dahl sheep hunters had some extra white gas that they did not want to haul out. Somehow, I was nominated and voted to go over to the Dahl sheep hunters' camp and beg for white gas. It was starting to get dark with grey skies and a cold breeze. As I approached their tent, I started to say "Hello" loudly and clapping my hands. You see, I did not want to get shot with them thinking I was bear! Finally, there was some rustling and talking in the tent, and four guys emerged from the tent. I mentioned that if they had some white gas that they did not want to haul out, we would like to pad our supply. The guide quickly and happily retrieved a half gallon of white gas and asked if we needed more food or other things. I said, "Probably not as we already had a lot of gear and our boats are probably going to be overloaded as it is." "How about some Dahl sheep meat? Would you like some of that?" asked the guide. "Nah," I said. "We do not have room." I thanked them for the white gas and returned to our camp. When I retold what had transpired, Neal, Karen, and Torre were all dumbstruck that I had refused Dahl sheep meat!! Neal volunteered to go back to the sheep hunters to see if he could take some of that Dahl sheep meat off their hands. Neal returned with a very big grin on his face and a couple of hands full of meat. I have to admit it was tasty and, fortunately, it was all gone before we loaded our boats the next morning to float northward. The group teased me about refusing Dahl sheep meat.

The next morning, we packed up our gear and inflated the boats. I was seriously skeptical of how that mountain of gear was going to fit into the boats. Torre's inflatable canoe was much smaller than our rubber raft / canoe hybrid, so I knew we would be carrying the bulk of the gear. Torre separated the gear into two piles, one pile for each boat. Neal and I struggled to get all the gear in our pile in our boat. Most of the gear went behind the center seat for the operator of oars. We wanted to keep the bow, or front of the boat, lighter to facilitate turning to avoid rocks. I have been kayaking and canoeing since high school and have built two kayaks. Neal had done canoe trips up in the Boundary Waters Canoe Area in northern Minnesota. Neal wanted the back seat and figured he would learn the oars on the fly. I would be up front with a canoe paddle. One trouble was that we had a rapid to run right away after launching from our camp. We got about a third of the way through the rapid, which curved around a bend in the river. Neal was having a catastrophic failure with the oars. Because our rubber raft was long and skinny, it made it hard to turn from the center of the boat. Furthermore, the long oar handles overlapped each other near the center of the boat. This meant that your oar hands were constantly colliding with each other unless you staggered your right and left oar strokes or made one oar lower in the water than the other. This was a lot to try and learn on the fly in a rapid. We were doing 180s through that rapid and hit both banks of the river at least twice. Fortunately, we did not capsize and did not damage the boat. Neal had had it with the boat. Either we would paddle it like or a canoe or I would run the oars. Since there was a long calm stretch of river in front of us, I agreed to give it a try. Unlike Neal, I had a still water initiation to the oars and how the boat handled. Our rubber boat was like a loaded semi-truck, while my kayak was more like a sports car. It took a lot of concentration and brain power to operate those overlapping oars. Plus, rather than face backward and pull on the oars, I wanted to see what was ahead and had to push forward on the oars for propulsion. So, it was kind of like doing sit-ups all day. I did OK on the first small rapid and improved as the day progressed. Sometimes I would have a mental lapse and instead of turning to avoid the rock, I would turn us right into the rock. Thankfully our boat was tough and forgiving. Neal and I learned on about the second day of boating that it also helped if we talked, so at least we were both trying to go on the same side of an obstacle. As day one of the river progressed, our boating confidence was building, but so was the difficulty of the rapids.

Given our remoteness and the impacts of losing gear or having team members get hypothermia after a swim in the cold river, Torre strove to err on the side of caution. We lined our boats through many rapids, with Neal handling the stern rope on our boat from shore while I waded in the river and guided our boat around obstacles. It was a cold, foggy day and, as always, we all had our chest waders on during boating days. Torre had warned us about tough rapids at the "S" curve part of the Hula Hula River. The "S" curve was near where the Hula Hula River emerged from the Brooks Range valley into the foothills. We were concentrating on lining our boats through the "S" curve rapids when Karen told me there were people here. That made no sense and I kind of ignored it. My focus was on our boats and the rapids. A couple of minutes later, a motion on the bank caught my eye and there stood a woman dressed in state-of-the-art outdoor clothes. Torre stopped our little train of boats and we tied them up to the thick willows on the bank. As we climbed up the bank through the willows, we broke out on the foggy tundra plain where two men were setting up a little camp stove. We learned that they were on a guided hiking trip. They had been dropped off by a small bush plane and were going to be picked up at some lake in another day or two to the west. Everyone seemed in good spirits, despite the fog and cold temperatures. You know you're in a remote location when you stop what you are doing and chat for 30 minutes when you see other human beings.

Once through the "S" curves, the rapids toned down a bit, but we still lined the boats through a couple of the rougher rapids. One stretch of the river went through a stretch with steep, rocky cliffs on both sides of the river. The cliffs may have towered 40 or 50 feet above the river. I was concentrating on avoiding another rock ahead when Neal pointed out a Dahl sheep on top of the cliffs. I watched as the sheep turned and disappeared into the ever-present fog. It was not long before Torre announced we had reached the spot where we would be camping. We set up camp and ate a hot supper in our "kitchen" tent, which was warmed by our two-burner stove that cooked our supper. The boating was much more physically demanding for me than the sampling days, so I slept well that night. One boating day my sleeping bag got damp in its dry bag in the boat. We heated water and poured it in my water bottle, put my water bottle in a sock, and threw it in the bottom of my sleeping bag. I stayed warm all night and by morning, my sleeping bag was dry.

The next sampling day was uneventful, but we covered a lot of ground. We summited the ridge that ran parallel to the river, about two miles from the river. We nonchalantly and confidently strode down into a saddle on the back side of the ridge and sampled a site. Neal was getting a soil description, measuring the thickness of the organic layer, and removing a soil plug for a picture of the soil profile. He was bent over a lot and working with things up close. I was walking around and getting the dominant species of vegetation, probing for permafrost, and getting GPS coordinates. It took us about 30 to 45 minutes per site. As we were finishing, Neal stood up and said, "Uh-oh, where's the river?" All the grassy slopes around us looked the same and he had lost his sense of direction working on his soil data collection. I said, "It's right over that rise," pointing westward on another foggy day. We strapped on our packs and walked up the rise and there was the river far below with our tents in the distance. Frankly, I too was relieved to see the river. If I had been wrong, what then? My GPS did not have the river in its map database and I had not added a waypoint at camp.

As the Hula Hula spilled out of the foothills and out onto the North Slope plains, rapids were less of an issue, but now it became a game of following the winding current as it zig-zagged through gravel bars. One lunch was on a cold foggy day and we just pulled out on this barren gravel bar. No break from the cold wind and we sat there eating our cold lunch. There was some scant conversation, but no whining.

On another sampling day, and this must have been fairly far south as there were still significant ridges distant from the river, Neal and I were sampling our way up the ridge. At a stop for me to catch my breath, Neal pointed out a herd of caribou on the other side of the river grazing their way toward us. We decided this was an ideal place for lunch and it was a sunny clear day. As we chomped on our lunch, we watched the herd swim across the river aways upstream from our camp. They then continued grazing their way straight up the ridge toward us. We got our cameras out, but were not quite sure what was going to happen. They walked right by us, maybe 20 to 30 feet away. They seemed curious and suspicious of us as we sat very still, just snapping pictures. The caribou continued on over the ridge top.

The next day, we were also going sampling, but this time on the east side of the river. There had been a lot of wind that night and I had put motion detectors up on wobbly little willows. I had to keep having to get up in the night for what I thought were false alarms. Bears like to chew on rubber and our boats were made of rubber. It would have been a major problem getting out of there without the boats. Walking was slow and difficult even on what looked to be flat, level ground. Uneven ground was masked by the tundra vegetation making it look almost as flat as a mowed lawn. I finally just shut the motion detectors off, as I was tired of running around in the dark, cold wind in the middle of the night. After breakfast, we got in the boats, crossed the river, tied the boats up, and clambered up a four-foot vertical dirt bank. Neal was ahead of me, and I was just clambering up on all fours trying to keep the shotgun from pointing in anyone's direction, when a moderate-sized, reddish brown animal ran off from just 10 to 15 feet away. It was a wolverine! I was relieved it ran away, as I was hardly in a position to do rapid, accurate shooting. I was ecstatic as I was now one of the few folks who has ever see a wolverine in the wild. It appeared to me he had been hiding where he could watch our camp across the river. Had the wolverine paid us a night visit and set my motion detectors off in the night?

After another long day in the boats, we were on a schedule to get to our Arctic Ocean rendezvous, and we had to put miles behind us as we moved north. We must have done a lot of walking to get the boats to float in shallow stretches, as Neal, Karen, and I were huddled around a willow fire on a gravel bar next to the river. We all had wet socks on our hands, which we held up by the fire to dry. We were chatting it up about something trivial and MAYBE a bit humorous. Torre was over in the "kitchen" tent talking to his wife on the satellite phone and maybe starting supper. Neal looked up and said, "There is a bear right across the river!" and pointed a little downstream. Then he said, "... and two more on top of the ridge right across from us." With that Neal disappeared to alert Torre, who had to say good-bye to his wife after she just overheard Neal telling Torre there were three grizzlies across the river. Comforting, I am sure, for a mother to hear that when her daughter is out there with us as well.

I really did not pay any attention to what Neal was doing. As I took two steps to get the shotgun in front of my tent, I was thinking, "This does not compute! Grizzlies are solitary animals! They are not supposed to do this!", and "Why is Neal always the first to see the bears?" Karen was standing a couple of feet toward the river from the fire. I stepped forward of her with my gun in the low ready position (barrel pointing down at about 45 degrees), finger on the slide release so I could rapidly rack in a live slug to the chamber. As Karen and I stood there in silence, the bear on the gravel bar across the river ran away into the bushes, and one of the two bears on top of the ridge ran off over the ridge. They must have seen us the same time we saw them. But the third one? He charged straight down that ridge toward Karen and me. It was kind of beautiful seeing the Grizzly's muscles rippling under its coat of fur and how quickly he (she?) covered the distance toward us. OK, that adoration was only a fleeting nanosecond. I set my perimeter distance where, if the bear crossed it, I was unloading four slugs into its chest. That was where the willows ended and gravel bar began on the other side of the river, a distance of 20 to 30 yards or so. I wanted to be able to get a couple of slugs in him (her) before it got into the water. Another thought was, what a pain it would be if we had a dead bear. You have to call the FWS or highway patrol, fill out a self-defense form, skin the bear, and turn the hide and head into the FWS. A bloody, heavy mess that would also attract more bears to our camp! Remarkably, the bear finally did a 180 and turned and ran back up and over the hill! I am not sure what the bear would have done if Karen had run, though. I was glad Karen had remained calm and stood her ground!

About then Torre and Neal rushed up from the "kitchen." Maybe their hurried arrival made the bear do the 180 turn? Safety in numbers maybe, as bears prefer to attack solitary or at least small groups? Anyway, Torre had me shoot the ridge they had gone over to make the bears think we were "bear hunters." In the rather intensive firearm training and annual recertification to prevent wild animal attacks both Neal and I had taken, they did not recommend warning shots. It was recommended to save all your rounds in the magazine for "the last Alamo." However, in this case, I had time to get a warning shot off, rack another round into the chamber, and reload the magazine. Extra rounds were accessible in my jacket pocket, and we practiced reloading while keeping an eye and gun on the target. I suspect a warning shot would have ended the bluff charge much sooner. Hindsight 20-20. All that matters is that we all lived.

Two days later, we were finishing up another long day in the boats. Torre and Karen had already landed their boats and were standing on the shore of our new campsite. Neal was in front and yelled, let's land just upstream from a big round rock poking out of the water about a foot. Neal was paddling furiously in the bow with the canoe paddle. I tried to swing the boat in that direction and give it more forward momentum, but to no avail. Either the current was faster than we thought, or my arms were wimpy after a long day at the oars. The bow of our boat, where Neal sat, impacted the upstream third of the rock, and the current immediately carried the rest of the boat downstream, kind of using the rock as pivot point. I was flailing with oars because if the oars dug into the gravel bottom and the current was pushing the boat in that direction, either the paddle would break or the boat would flip. Somehow, Neal was now standing on the rock, for an instant. He lost his balance, fell on his back into the one- or two-foot deep COLD water, and then appeared to stand on his head for an instant as he tried to stand back up quickly.

I felt bad and was worried about Neal getting hypothermia, but he claimed he was OK. I think it was about 30 minutes later he asked me for my version of what had happened. I guess the moral of the story is: 1) do not overestimate the strength of the old guy's arms after a long day of paddling, 2) be careful landing upstream from a rock, and 3) damn, that water is COLD!!

Another time, after a long day of sampling (Neal and I had gone a long ways from camp that day, up over the ridge and down into the next valley), after supper we were thinking of hitting the sack. Either Karen or Neal spotted a grizzly munching away on blueberries about 300 or 400 yards away. The bear was slowly grazing northward along the base of the ridge as the river valley was getting broader as we progressed northward. We watched the bear for 40 minutes or so while we were getting ready for bed and doing the dishes. The bear was oblivious to our being there and in no hurry at all. Certainly, going to bed with a bear in sight sounded like a bad option. It was still pretty light out, but we were exhausted and the next day was to be another long boating day.

Torre suggested that I get one of my flare rounds and bring my shotgun. Torre and I were going to go scare the bear off. The flare does pose the potential of starting a fire, but the area where the flare would probably fall was a soupy tundra with water at or just above the surface of ground. Torre had a pistol, which I think was a 44 mag. Torre and I walked about 30 yards out of camp toward the bear and stopped. Torre fired his pistol into the air. The bear stopped eating and sat up on its rear haunches staring at us. Torre said, "Shoot the flare!" The flare round looked like a regular 12-gauge round, except it had a large nose cone on it kind of like the plastic bird cage like thing they used to put on broken fingers. I think this was to prevent the flares being fired in rapid succession out of a shotgun. With difficulty, I finally got the flare into the chamber, aimed at the bear, and then elevated my aim about 40 degrees upward. When I pulled the trigger, we could see the flare arcing upward and the bear just sitting there watching. I guess a warning shot would not have worked on this bear. But when the flare started to arc downward toward the bear, it took off at a full run up the ridge. As I watched the bear run, I was also trying to eject the spent flare round from the chamber. This is standard operating procedure for me; that is, get a live round in there as quick as possible after a warning shot or a bean bag shot just in case the bear charges instead of running off. To my horror, the long nose piece hampered the ejection of the spent flare, and when I racked the fore stock of the shotgun, it brought in an additional live round. Now I had both a live round and a spent flare all in that central bolt and ejection port area. I wrestled with it for about a minute before finally getting the spent flare out. I was glad the bear had not charged! I replaced the slug in the chamber with a bird shot round (I had a few of them along in case we had to go into survival mode for Ptarmigan, water fowl, or marmots to eat). With all this bear activity, I wanted to save my slug rounds. We were unsure what the situation would be when we got into polar bear (carnivorous not omnivorous like the grizzly) country near the Arctic Ocean. Torre and I watched that bear running up the ridge after what I estimated to be a mile, then it stopped and looked back. That is when I fired the bird shot and the bear took off running again. When Torre and I returned to camp, Neal pointed out the bear to me. It was grazing near the top of the ridge about 2 miles away. We went to bed anyway.

As we got closer to the ocean, the terrain got flatter. This boating day was another foggy, misty day. As Neal and I were paddling along, Neal said, "Be real quiet as there is a big bear near the shore who has not seen us." This was a little nerve-racking, as the shotgun was in a soft case to protect it from the water. It would have taken at least a minute to get a shot off. If that bear charged, however, we would not have a minute. I took care not to splash my oars and just stared at the bear as we glided by. Mr. Bear was busy eating berries, trying to put on enough fat to carry it through the winter.

Mid-afternoon, Torre and Karen pulled out on the eastern bank. Torre announced that we were very close to the GPS portage location. Here was where we needed to portage over to the Okpilak River. I would say the distance was 200-300 yards, but across soggy tundra, which Neal pointed out were actually low center permafrost polygons. As I was getting ready to take a load of gear and head across to the Okpilak, Neal spotted yet another grizzly off in the distance. So, I got to stand guard with the shotgun while the rest of the team made multiple trips. After a while, the bear moved on, and I helped carry gear, making three or four trips back and forth. Torre told us that we were definitely in polar bear country now, so be on the lookout. We set up our tents on the edge of the Okpilak. Torre was concerned that my tent was a bit far from the others, but it did not strike me that it was that much of an outlier. Anyway, it was three-fourths set up, so it stayed. I had suggested that a weather report would be good, as I did not want to be boating in the Arctic Ocean when there was a lot of wind and waves. Torre called on his sat phone and learned that 40-knot winds were forecast for tomorrow, which was to be a sampling day. The following day, which would be our rendezvous with the boat from Kaktovik in the Avery Island lagoon, would be sunny and calm! Torre let me talk to Bruce in Kaktovik and confirm our pickup. I asked about polar bears and he suggested a warning shot usually worked well. We readied our camp for the 40-knot winds that were coming the next day. Oars and paddles were stuck in the ground and used as additional tent staking. I had one paddle for my tent, a Eureka three-man pup tent with an external frame, as Torre thought my tent would not stand up to the wind. I had a few longer stakes that I used in critical positions. Everyone else had small dome tents.

The next morning the winds were getting strong. After breakfast, Torre warned us about losing our data sheets to the wind. I took note and moved the other day's data sheets out of my clipboard and into my camping backpack. Neal and I were out sampling in the permafrost polygons in the strong wind, struggling to just stand, when Neal said it looked like our "kitchen" tent was "screwed up." We headed back to camp after finishing that site and found the large dome "kitchen" tent pretty much prone to the ground. We removed the pole ends from their moorings to relieve tension and let the tent lay flat on the ground. We put some gear on top of the tent to keep it from blowing away. Neal's dome tent was flattened as well, but the rest of the tents were still standing. Soon Torre and Karen returned. They had sampled down toward the coast, also scouting our route for tomorrow down the Okpilak. But they turned back when a couple of white dots on a small hill up ahead kept moving around. We decided about all we could do was take a nap. So, we all climbed into our tents for a while. Neal just laid in his prone tent. After a while the wind subsided for a bit, and I could hear Karen rustling in her tent, so I climbed out of mine. Soon all four of us were up and proceeded to try and resurrect the "kitchen." Somehow, we were able to get the big tent back up—a miracle! Soon supper was being prepared. Neal decided he was going to sleep in the "kitchen" that night since his tent had collapsed. I asked if he wanted the shotgun, and he said "Yes." The liabilities of sleeping in the kitchen were that it had food, which is a smell that bears like. Also, you could probably not sleep in as easily because there was an early riser in the group (me).

Morning came, and the sun was out and the wind gone! This was to be our last boating day! There was some trepidation about being out on the Arctic Ocean (swimming polar bears, tidal currents, and no wind protection). Would we be able to float our boats through the Okpilak's delta, and what about polar bears on land? We got the boats loaded, and Neal agreed to give the oars a go. The banks of the Okpilak were near vertical and about chest high, so going down the river in our boats we could not see around us very well. This made me nervous about a polar bear surprising us and peering down on us over the edge of the steep bank. As a result, the shotgun was out of the scabbard and between my legs, barrel down, as I paddled with the canoe paddle. Neal had turned the boat 180 degrees so he could get more leverage pulling on the oars. This put me at the stern with the canoe paddle, where most of the steering was done. The river was very braided and shallow, knee deep at the most. Those two white moving dots that Torre and Karen had seen yesterday were still there on the small hill along the river up ahead. We had little option but to follow the river toward the white dots. As we approached the mother and cub polar bears, they were watching us. Mom actually started walking out across the mudflats toward us. I asked Torre if I should do a warning shot and he said, "No." The polar bear is endangered, and the FWS had given us strict guidance to, in particular, give mothers with cubs a wide berth and not to harass them. I was getting real nervous, as momma bear approached with baby in tow. Midway across the mudflat, she recognized us as human and turned and ran.

The big sand bar at the mouth of the Okpilak was just over ankle deep. We could walk and pull our boat along fine, but if we jumped in, the boat would drag bottom. As we got further out, gradually the water depth increased, and we were able to jump into the boats. We headed northeast and across the lagoon, landing at Avery Island to sample our last site. This completed our gradient from the mountains to the ocean. Karen took a picture of Neal, Torre, and me standing in the ocean on the Arctic Ocean side (north side) of Avery Island. There was not much on Avery Island besides a lot of driftwood and sand.

Back in the boats we headed east. Somewhere out here we were supposed to meet Bruce from Kaktovik. Finally, someone spotted a motorboat headed our way. Once Bruce got close, he cut his engine and waited for us to paddle out to him. He did not want to get his boat in too shallow of water. We passed our gear to Bruce and then climbed aboard his boat. We then pulled our boats in and set them across the back of Bruce's boat. We huddled in the cab of Bruce's boat on the ride back to Kaktovik. Bruce dumped us out on a beach near the airport and said he would call our hotel to come pick us up. We deflated our boats and crudely bundled them up before the old school bus from the hotel showed up. I was really appreciating Bruce's twin outboard boat and the warm bus ride back to the hotel. Our hotel did not look very fancy. It looked like a double wide trailer with a shipping container attached, but it was cozy and warm inside and the food was included! All you could eat! The hotel had a large group of fun-loving, elderly women from Australia and another middle-aged fellow from Norway. They were all here to see the polar bears. It turns out that Kaktovik is one of the best places in the world to see polar bears. Bruce had told us on the boat ride in that they had harvested a whale two days prior to our arrival. A fistful of polar bears was on the whale bone pile, which could be seen through the spotting scope at the hotel window. Karen and I went out on the hotel's bus to park next to the bone pile for about 45 minutes. I counted 38 polar bears, with five new ones just swimming in the ocean and climbing out to feed on the bone pile while we were there. The Australian women thought the polar bears were cute. I suggested they step outside the bus and say that. I told them I had been camping in a tent not far from here last night and that the bears looked foreboding to me.

Torre had some data loggers that he needed to collect that measured soil temperatures over the summer. Neal went with Torre to help, taking the shotgun along. Karen and I worked to pack up the gear, as the hotel wanted to weigh our bags and get the baggage fees paid before our commercial flight arrived midday. My fingers were cracked and bleeding from working in the cold weather. One should bring bag balm for cow's udders, apply it to your hands nightly, then sleep with gloves on to prevent this. That is what the guys at a bar in Fairbanks told me later. Every time I bumped my finger on the gear or had to open or close a zipper, they would hurt badly and start to bleed. Nonetheless, Karen and I got all the gear ready to ship and weighed. Torre and Neal got back just as the bus was about to leave for the airport. We helped load our bags into our commercial plane back to Fairbanks.

Torre was on a different flight home, as he had more data loggers to retrieve at Dead Horse. Our flight home was cloudy and foggy, and I was so tired and relieved that all our sequential dependencies had successfully delivered us back into civilization, that I fell asleep on the plane. I was snoring, and Neal took a picture of me sleeping. As we climbed over the Brooks Range, I woke and soon had my nose pressed to the window. As we reached the southern edge of the Brooks Range, the weather changed from cloudy and foggy to clear and sunny. Two totally different air masses, with maritime influences on the north side for the Brooks, and more continental air masses in the interior. We landed in Arctic Village along the Chandalar River and had to wait for a student going to the University of Alaska at Fairbanks. I guess he thought his flight was tomorrow. Someone had to go get him at home on a four-wheeler while we waited. But no hurry from us, we were pretty content and considered this the lap of luxury.

At the Fairbanks airport, we claimed our mountain of baggage. Torre's wife came and picked up Karen. Neal volunteered to go to the USGS office and get the Suburban to pick me up and the gear. Meanwhile, I sat at the airport guarding the pile of gear, boats and all. I was quite a spectacle, as several folks asked where we had been. It took Neal a long time to return, and I was starting to get a bit concerned, but he did finally show up. It turned out the battery on the Suburban was dead, and he had to charge the battery a bit before he could start it. One always needs a little drama I guess.

This Hula Hula trip was definitely a "bucket list" trip. We saw 12 grizzlies, 41 polar bears, 1 wolverine, and oodles of caribou. Karen and Torre saw the wolf.

This compilation is only about 30% of my nonfiction campfire stories. More to come if this book reviews well.